SpringerBriefs in Mathematics

SpringerBriefs in Mathematics showcases expositions in all areas of mathematics and applied mathematics. Manuscripts presenting new results or a single new result in a classical field, new field, or an emerging topic, applications, or bridges between new results and already published works, are encouraged. The series is intended for mathematicians and applied mathematicians.

Titles from this series are indexed by Web of Science, Mathematical Reviews, and zbMATH.

More information about this series at http://www.springer.com/series/10030

Clara Löh

Ergodic Theoretic Methods in Group Homology

A Minicourse on L^2-Betti Numbers in Group Theory

 Springer

Clara Löh
Fakultät für Mathematik
Universität Regensburg
Regensburg, Bayern, Germany

ISSN 2191-8198   ISSN 2191-8201   (electronic)
SpringerBriefs in Mathematics
ISBN 978-3-030-44219-4   ISBN 978-3-030-44220-0   (eBook)
https://doi.org/10.1007/978-3-030-44220-0

Mathematics Subject Classification (2020): 55N25, 20J05, 37A20, 57N65, 22E40, 20P05, 46L10

This Springer imprint is published by the registered company Springer Nature Switzerland AG
The registered company address is: Gewerbestrasse 11, 6330 Cham, Switzerland

für A ∗ A

Contents

0   Introduction   1

1   The von Neumann dimension   5

   1.1   From the group ring to the group von Neumann algebra   6

        1.1.1   The group ring   6

        1.1.2   Hilbert modules   8

        1.1.3   The group von Neumann algebra   9

   1.2   The von Neumann dimension   11

   1.E   Exercises   15

2 L^2-Betti numbers   17

   2.1   An elementary definition of L^2-Betti numbers   18

        2.1.1   Finite type   18

        2.1.2 L^2-Betti numbers of spaces   19

        2.1.3 L^2-Betti numbers of groups   20

   2.2   Basic computations   20

        2.2.1   Basic properties   21

        2.2.2   First examples   23

   2.3   Variations and extensions   24

   2.E   Exercises   26

3   The residually finite view: Approximation   27

   3.1   The approximation theorem   28

   3.2   Proof of the approximation theorem   28

        3.2.1   Reduction to kernels of self-adjoint operators   29

        3.2.2   Reformulation via spectral measures   29

        3.2.3   Weak convergence of spectral measures   30

        3.2.4   Convergence at 0   31

3.3		Homological gradient invariants	32
	3.3.1	Betti number gradients	33
	3.3.2	Rank gradient	33
	3.3.3	More gradients	34
3.E		Exercises	35

4 The dynamical view: Measured group theory — 37

4.1		Measured group theory	38
	4.1.1	Standard actions	38
	4.1.2	Measure/orbit equivalence	39
4.2		L^2-Betti numbers of equivalence relations	40
	4.2.1	Measured equivalence relations	41
	4.2.2	L^2-Betti numbers of equivalence relations	41
	4.2.3	Comparison with L^2-Betti numbers of groups	43
	4.2.4	Applications to orbit equivalence	46
	4.2.5	Applications to L^2-Betti numbers of groups	46
4.3		Cost of groups	47
	4.3.1	Rank gradients via cost	48
	4.3.2	The cost estimate for the first L^2-Betti number	52
	4.3.3	Fixed price	55
4.E		Exercises	57

5 Invariant random subgroups — 59

5.1		Generalised approximation for lattices	60
	5.1.1	Statement of the approximation theorem	60
	5.1.2	Terminology	61
5.2		Two instructive examples	62
	5.2.1	Lattices in $\mathrm{SL}(n, \mathbb{R})$	62
	5.2.2	Why doesn't it work in rank 1 ?!	63
5.3		Convergence via invariant random subgroups	64
	5.3.1	Invariant random subgroups	64
	5.3.2	Benjamini–Schramm convergence	66
	5.3.3	The accumulation point	68
	5.3.4	Reduction to Plancherel measures	69
	5.3.5	Convergence of Plancherel measures	69
5.E		Exercises	71

6 Simplicial volume — 73

6.1		Simplicial volume	74
6.2		The residually finite view	75
6.3		The dynamical view	77
6.4		Basic proof techniques	79
	6.4.1	The role of the profinite completion	79
	6.4.2	Betti number estimates	81
	6.4.3	The rank gradient/cost estimate	83
	6.4.4	Amenable fundamental group	84
	6.4.5	Hyperbolic 3-manifolds	87
	6.4.6	Aspherical 3-manifolds	90
6.E		Exercises	92

Contents

A Quick reference   93
   A.1 Von Neumann algebras   94
   A.2 Weak convergence of measures   94
   A.3 Lattices   95

Bibliography   97

Symbols   109

Index   111

0

Introduction

This is an extended version of the lecture notes for a five lecture mini-course at the MSRI summer graduate school *Random and arithmetic structures in topology* (organised by Alexander Furman and Tsachik Gelander) in June 2019.

This minicourse gives a brief introduction to ergodic theoretic methods in group homology. By now, this is a vast subject [62]. In the present course, we will focus on L^2-Betti numbers.

The underlying fundamental observation is that taking suitable coefficients for group (co)homology allows us to connect homological invariants with ergodic theory; good candidates are coefficients related to non-commutative measure theory, i.e., to von Neumann algebras and dynamical systems. This interaction works in both directions:

- On the one hand, (co)homology with coefficients based on dynamical systems leads to orbit/measure equivalence invariants.

- On the other hand, many homological gradient invariants can be accessed through the dynamical system given by the profinite completion.

Overview of this minicourse

- **Chapter 1.** We will start our gentle introduction to L^2-Betti numbers by introducing the von Neumann dimension. In order to keep the technical overhead at a minimum, we will work with an elementary approach to von Neumann dimensions and L^2-Betti numbers.

© The Author(s), under exclusive license to Springer Nature Switzerland AG 2020
C. Löh, *Ergodic Theoretic Methods in Group Homology*,
SpringerBriefs in Mathematics, https://doi.org/10.1007/978-3-030-44220-0_1

- **Chapter 2.** The von Neumann dimension allows us to define L^2-Betti numbers. We will explore basic computational tools and calculate L^2-Betti numbers in simple examples.

- **Chapter 3.** On the one hand, L^2-Betti numbers are related to classical Betti numbers through approximation. This residually finite view has applications to homological gradient invariants.

- **Chapter 4.** On the other hand, L^2-Betti numbers are also related to measured group theory. We will compare this dynamical view with the residually finite view.

- **Chapter 5.** Via invariant random subgroups, ergodic theory also gives a new way of obtaining approximation results for normalised Betti numbers of lattices/locally symmetric spaces.

- **Chapter 6.** Finally, we will use the dynamical approach to L^2-Betti numbers and related invariants as a blueprint to prove approximation results for simplicial volume.

None of the material in this book is original; in particular, Chapters 1–3 are covered thoroughly in textbooks on L^2-Betti invariants [105, 84] and large parts of Chapter 4 can be found in surveys on measured group theory [65, 62].

Prerequisites. I tried to keep things as elementary as reasonably possible; this means that a basic background in algebraic topology (fundamental group, covering theory, (co)homology, classical group (co)homology), functional analysis (bounded operators, measure theory), and elementary homological algebra should be sufficient to follow most of the topics of the book. Chapter 5 requires some knowledge on lattices/locally symmetric spaces and Chapter 6 requires moderate familiarity with the topology and geometry of manifolds.

Some basic notions on von Neumann algebras, weak convergence of measures, and lattices are collected in Appendix A.

Exercises. Each chapter ends with a small selection of exercises; moreover, small on-the-fly exercises are marked with "(check!)" in the text.

Conventions. The set $\mathbb{N}$ of natural numbers contains 0. All rings are unital and associative (but very often *not* commutative). By default, modules are left modules (if not specified otherwise).

Additional material

This brief introduction is in no way exhaustive, but covers only a few selected topics. I hope that these notes inspire the reader to get involved with more systematic treatments and the original research literature.

Textbooks on L^2-invariants
- H. Kammeyer. *Introduction to ℓ^2-invariants*, Springer Lecture Notes in Mathematics, 2247, 2019.
- W. Lück. *L^2-Invariants: Theory and Applications to Geometry and K-Theory*, Ergebnisse der Mathematik und ihrer Grenzgebiete, 44, Springer, 2002.

Recommended further reading
- M. Abért, N. Bergeron, I. Biringer, T. Gelander, N. Nikolov, J. Raimbault, I. Samet. On the growth of L^2-invariants for sequences of lattices in Lie groups, *Ann. of Math. (2)*, 185(3), pp. 711–790, 2017.
- M. Abért, N. Nikolov. Rank gradient, cost of groups and the rank versus Heegaard genus problem, *J. Eur. Math. Soc.*, 14, 1657–1677, 2012.
- M. W. Davis. *The Geometry and Topology of Coxeter Groups*, London Mathematical Society Monographs, 32, Princeton University Press, 2008.
- B. Eckmann. Introduction to l_2-methods in topology: reduced l_2-homology, harmonic chains, l_2-Betti numbers, notes prepared by Guido Mislin, *Israel J. Math.*, 117, pp. 183–219, 2000.
- A. Furman. A survey of measured group theory. In *Geometry, Rigidity, and Group Actions* (B. Farb, D. Fisher, eds.), 296–347, The University of Chicago Press, 2011.
- D. Gaboriau. Coût des relations d'équivalence et des groupes, *Invent. Math.*, 139(1), 41–98, 2000.
- D. Gaboriau. Invariants ℓ^2 de relations d'équivalence et de groupes, *Inst. Hautes Études Sci. Publ. Math.*, 95, 93–150, 2002.
- T. Gelander. A view on invariant random subgroups and lattices, *Proceedings of the International Congress of Mathematicians. Volume II*, pp. 1321–1344, World Sci. Publ., 2018.
- A. S. Kechris, B. D. Miller. *Topics in Orbit Equivalence*, Springer Lecture Notes in Mathematics, 1852, 2004.
- D. Kerr, H. Li. *Ergodic theory. Independence and dichotomies*, Springer Monographs in Mathematics, Springer, 2016.
- W. Lück. Approximating L^2-invariants by their finite-dimensional analogues, *Geom. Funct. Anal.*, 4(4), pp. 455–481, 1994.
- J. Raimbault. Blog, https://perso.math.univ-toulouse.fr/jraimbau/
- R. Sauer. Amenable covers, volume and L^2-Betti numbers of aspherical manifolds, *J. Reine Angew. Math.*, 636, 47–92, 2009.

Videos. The videos of the original lectures within the MSRI summer graduate school are available at https://www.msri.org/people/23160 . Moreover, material for the other two minicourses of this summer school is available at https://www.msri.org/summer_schools/853 .

Errata. Comments and corrections for these notes can be submitted by email to clara.loeh@mathematik.uni-r.de; errata will be collected at

http://www.mathematik.uni-r.de/loeh/l2_book/errata.pdf .

Acknowledgements. I would like to thank Alex Furman and Tsachik Gelander for organising the summer school and inviting me as a lecturer. Special thanks go to Gil Goffer, who skillfully taught the discussion sessions for this course.

Moreover, I am grateful to Roman Sauer for sharing his notes on Chapter 3, to Damien Gaboriau for valuable comments on Chapter 4, to Miklos Abért for a lesson on IRS in the Schwarzwald, and to Marco Moraschini for patient proof-reading.

This book project has been partially supported by the MSRI (through funding of the summer school) and the SFB 1085 Higher Invariants (funded by the DFG, Universität Regensburg).

Regensburg, February 2020   Clara Löh

1

The von Neumann dimension

Betti numbers are dimensions of (co)homology groups. In the presence of a group action, we can alternatively also use an equivariant version of dimension; this leads to L^2-Betti numbers.

In this chapter, we will introduce such an equivariant version of dimension, using the group von Neumann algebra. In Chapter 2, this dimension will allow us to define L^2-Betti numbers of groups and spaces.

Overview of this chapter.

1.1	From the group ring to the group von Neumann algebra	6
1.2	The von Neumann dimension	11
1.E	Exercises	15

Running example. the additive group $\mathbb{Z}$, finite (index sub)groups

© The Author(s), under exclusive license to Springer Nature Switzerland AG 2020
C. Löh, *Ergodic Theoretic Methods in Group Homology*,
SpringerBriefs in Mathematics, https://doi.org/10.1007/978-3-030-44220-0_2

1.1 From the group ring
to the group von Neumann algebra

Examples of fundamental invariants in algebraic topology are Betti numbers of spaces and groups, which are numerical invariants, extracted from (co)homology by taking dimensions of homology groups.

We will now pass to an equivariant setting: Let $\Gamma \curvearrowright X$ be a continuous group action on a topological space X. Then the singular chain complex $C_*^{\mathrm{sing}}(X; \mathbb{C})$ and the singular homology $H_*(X; \mathbb{C})$ of X inherit a Γ-action, and thus consist of modules over the group ring $\mathbb{C}\Gamma$.

Unfortunately, the group ring $\mathbb{C}\Gamma$, in general, does not admit an accessible module/dimension theory. We will therefore pass to completions of the group ring: $\ell^2\Gamma$ (for the modules) and the von Neumann algebra $N\Gamma$ (for the morphisms), which lead to an appropriate notion of traces and thus to a Γ-dimension. We will now explain this in more detail.

1.1.1 The group ring

The group ring of a group Γ is an extension of the ring $\mathbb{C}$ with new units coming from the group Γ:

Definition 1.1.1 (group ring). Let Γ be a group. The *(complex) group ring* of Γ is the $\mathbb{C}$-algebra $\mathbb{C}\Gamma$ (sometimes also denoted by $\mathbb{C}[\Gamma]$ to avoid misunderstandings)

- whose underlying $\mathbb{C}$-vector space is $\bigoplus_{g \in \Gamma} \mathbb{C}$, freely generated by Γ (we denote the basis element corresponding to $g \in \Gamma$ simply by g),

- and whose multiplication is the $\mathbb{C}$-bilinear extension of composition in Γ, i.e.:

$$\cdot : \mathbb{C}\Gamma \times \mathbb{C}\Gamma \longrightarrow \mathbb{C}\Gamma$$

$$\left(\sum_{g \in \Gamma} a_g \cdot g, \sum_{g \in \Gamma} b_g \cdot g \right) \longmapsto \sum_{g \in \Gamma} \sum_{h \in \Gamma} a_h \cdot b_{h^{-1} \cdot g} \cdot g$$

(where all sums are "finite", i.e., all but finitely many coefficients are 0).

Example 1.1.2 (group rings).

- The group ring of "the" trivial group 1 is just $\mathbb{C}[1] \cong_{\mathrm{Ring}} \mathbb{C}$.

- The group ring $\mathbb{C}[\mathbb{Z}]$ of the additive group $\mathbb{Z}$ is isomorphic to $\mathbb{C}[t, t^{-1}]$, the ring of Laurent polynomials over $\mathbb{C}$ (check!).

- Let $n \in \mathbb{N}_{>0}$. Then we have $\mathbb{C}[\mathbb{Z}/n] \cong_{\text{Ring}} \mathbb{C}[t]/(t^n - 1)$ (check!).

- In general, group rings are *not* commutative. In fact, a group ring $\mathbb{C}\Gamma$ is commutative if and only if the group Γ is Abelian (check!). Hence, for example, the group ring $\mathbb{C}[F_2]$ of "the" free group F_2 of rank 2 is not commutative.

Caveat 1.1.3 (notation in group rings). When working with elements in group rings, some care is required. For example, the term $4 \cdot 2$ in $\mathbb{C}[\mathbb{Z}]$ might be interpreted in the following different ways:

- the product of 4 times the ring unit and 2 times the ring unit, or

- 4 times the *group* element 2.

We will circumvent this issue in $\mathbb{C}[\mathbb{Z}]$, by using the notation "t" for a generator of the additive group $\mathbb{Z}$ and viewing the infinite cyclic group $\mathbb{Z}$ as multiplicative group. Using this convention, the first interpretation would be written as $4 \cdot 2$ (which equals 8) and the second interpretation would be written as $4 \cdot t^2$. Similarly, also in group rings over other groups, we will try to avoid ambiguous notation.

Proposition 1.1.4 (group ring, universal property). *Let Γ be a group. Then the group ring $\mathbb{C}\Gamma$, together with the canonical inclusion map $i \colon \Gamma \longrightarrow \mathbb{C}\Gamma$ (as standard basis) has the following universal property: For every $\mathbb{C}$-algebra R and every group homomorphism $f \colon \Gamma \longrightarrow R^\times$, there exists a unique $\mathbb{C}$-algebra homomorphism $\mathbb{C}f \colon \mathbb{C}\Gamma \longrightarrow R$ with $\mathbb{C}f \circ i = f$.*

$$
\begin{array}{ccc}
\Gamma & \xrightarrow{\;f\;} R^\times \xrightarrow{\;\text{incl}\;} R \\
{\scriptstyle i}\big\downarrow & \nearrow \\
\mathbb{C}\Gamma & {\scriptstyle \exists! \; \mathbb{C}f}
\end{array}
$$

Proof. This is a straightforward calculation (check!).   □

Outlook 1.1.5 (Kaplansky conjecture). The ring structure of group rings is not well understood in full generality. For example, the following versions of the Kaplansky conjectures are still open: Let Γ be a torsion-free group.

- Then the group ring $\mathbb{C}\Gamma$ is a domain (?!).

- The group ring $\mathbb{C}\Gamma$ does not contain non-trivial idempotents (?!). (I.e., if $x \in \mathbb{C}\Gamma$ with $x^2 = x$, then $x = 1$ or $x = 0$).

However, a positive solution is known for many special cases of groups [38, 117, 44][105, Chapter 10] (such proofs often use input from functional analysis or geometry) and no counterexamples are known.

1.1.2 Hilbert modules

Homology modules are quotient modules. In the presence of an inner product, quotients of the form A/B (by closed subspaces B) can be viewed as *submodules* of A (via orthogonal complements). Therefore, we will pass from the group ring $\mathbb{C}\Gamma$ to the completion $\ell^2\Gamma$:

Definition 1.1.6 ($\ell^2\Gamma$). Let Γ be a group. Then

$$\langle\,\cdot\,,\,\cdot\,\rangle\colon \mathbb{C}\Gamma \times \mathbb{C}\Gamma \longrightarrow \mathbb{C}$$

$$\left(\sum_{g\in\Gamma} a_g \cdot g, \sum_{g\in\Gamma} b_g \cdot g\right) \longmapsto \sum_{g\in\Gamma} \overline{a}_g \cdot b_g$$

is an inner product on $\mathbb{C}\Gamma$. The completion of $\mathbb{C}\Gamma$ with respect to this inner product is denoted by $\ell^2\Gamma$ (which is a complex Hilbert space). More concretely, $\ell^2\Gamma$ is the $\mathbb{C}$-vector space of ℓ^2-summable functions $\Gamma \longrightarrow \mathbb{C}$ with the inner product

$$\langle\,\cdot\,,\,\cdot\,\rangle\colon \ell^2\Gamma \times \ell^2\Gamma \longrightarrow \mathbb{C}$$

$$\left(\sum_{g\in\Gamma} a_g \cdot g, \sum_{g\in\Gamma} b_g \cdot g\right) \longmapsto \sum_{g\in\Gamma} \overline{a}_g \cdot b_g.$$

Example 1.1.7.

- If Γ is a finite group, then $\ell^2\Gamma = \mathbb{C}\Gamma$.

- If $\Gamma = \mathbb{Z} = \langle t \mid \rangle$, then Fourier analysis shows that

$$F\colon \ell^2\Gamma \longrightarrow L^2\big([-\pi,\pi],\mathbb{C}\big)$$

$$\sum_{n\in\mathbb{Z}} a_n \cdot t^n \longmapsto \left(x \mapsto \frac{1}{\sqrt{2\pi}} \cdot \sum_{n\in\mathbb{Z}} a_n \cdot e^{i\cdot n\cdot x}\right)$$

is an isomorphism of $\mathbb{C}$-algebras (with inner product).

Remark 1.1.8 (countability and separability). In order to avoid technical complications, in the following, we will always work with countable groups; then, the associated ℓ^2-space will be separable.

Definition 1.1.9 (Hilbert modules). Let Γ be a countable group.

- A *Hilbert Γ-module* is a complex Hilbert space V with a $\mathbb{C}$-linear isometric (left) Γ action such that there exists an $n \in \mathbb{N}$ and an isometric Γ-embedding $V \longrightarrow (\ell^2\Gamma)^n$. Here, we view $\ell^2\Gamma$ as a left $\mathbb{C}\Gamma$-module via

$$\Gamma \times \ell^2\Gamma \longrightarrow \ell^2\Gamma$$

$$(g, f) \longmapsto (x \mapsto f(x \cdot g)).$$

- Let V and W be Hilbert Γ-modules. A *morphism* $V \longrightarrow W$ *of Hilbert Γ-modules* is a Γ-equivariant bounded $\mathbb{C}$-linear map $V \longrightarrow W$.

In a complex Hilbert space, we have the following fundamental equality for (closed) submodules A (check!):

$$\mathbb{C}\text{-dimension of } A = \text{trace of the orthogonal projection onto } A.$$

For Hilbert Γ-modules, we will use this description of the dimension as a *definition*. Therefore, we first need to be able to describe orthogonal projections and we need a suitable notion of trace. Both goals can be achieved by means of the group von Neumann algebra.

1.1.3 The group von Neumann algebra

Let Γ be a countable group and let $a \in \mathbb{C}\Gamma$. Then the right multiplication map $M_a \colon \ell^2\Gamma \longrightarrow \ell^2\Gamma$ by a is a (left) Γ-equivariant isometric $\mathbb{C}$-linear map. Similarly, matrices A over $\mathbb{C}\Gamma$ induce morphisms M_A between finitely generated free $\ell^2\Gamma$-modules.

However, morphisms of Hilbert Γ-modules, in general, will not be of this simple form: We will need more general matrix coefficients.

Definition 1.1.10 (group von Neumann algebra). Let Γ be a countable group.

- Let $B(\ell^2\Gamma)$ be the $\mathbb{C}$-algebra of bounded linear operators $\ell^2\Gamma \longrightarrow \ell^2\Gamma$.

- The *group von Neumann algebra of* Γ is the weak closure of $\mathbb{C}\Gamma$ (acting by right multiplication on $\ell^2\Gamma$) in $B(\ell^2\Gamma)$.

Remark 1.1.11 (alternative descriptions of the group von Neumann algebra). Let Γ be a countable group. Then the group von Neumann algebra $N\Gamma$ is a von Neumann algebra (Definition A.1.1) and thus can equivalently be described as follows (Theorem A.1.2):

- $N\Gamma$ is the strong closure of $\mathbb{C}\Gamma$ (acting by right multiplication on $\ell^2\Gamma$).

- $N\Gamma$ is the bicommutant of $\mathbb{C}\Gamma$ (acting by right multiplication on $\ell^2\Gamma$).

- $N\Gamma$ is the subalgebra of $B(\ell^2\Gamma)$ consisting of all bounded operators that are *left* $\mathbb{C}\Gamma$-equivariant.

Theorem 1.1.12 (von Neumann trace). *Let* Γ *be a countable group and let*

$$\operatorname{tr}_\Gamma \colon N\Gamma \longrightarrow \mathbb{C}$$
$$a \longmapsto \langle e, a(e)\rangle,$$

where $e \in \mathbb{C}\Gamma \subset \ell^2\Gamma$ *denotes the atomic function at* $e \in \Gamma$. *Then* tr_Γ *satisfies the following properties:*

1. *Trace property. For all $a,b \in N\Gamma$, we have $\mathrm{tr}_\Gamma(a \circ b) = \mathrm{tr}_\Gamma(b \circ a)$.*

2. *Faithfulness. For all $a \in N\Gamma$, we have $\mathrm{tr}_\Gamma(a^* \circ a) = 0$ if and only if $a = 0$. Here, a^* denotes the adjoint operator of a.*

3. *Positivity. For all $a \in N\Gamma$ with $a \geq 0$, we have $\mathrm{tr}_\Gamma a \geq 0$. Here, $a \geq 0$ if and only if $\langle x, a(x)\rangle \geq 0$ for all $x \in \ell^2\Gamma$.*

Proof. *Ad 1.* A straightforward computation shows that the trace property holds on the subalgebra $\mathbb{C}\Gamma$ (check!). By construction, tr_Γ is weakly continuous. Therefore, the trace property also holds on $N\Gamma$.

Ad 2. For the non-trivial implication, let $a \in N\Gamma$ with $\mathrm{tr}_\Gamma(a^* \circ a) = 0$. Then, by definition, we have

$$0 = \mathrm{tr}_\Gamma(a^* \circ a) = \langle e, a^* \circ a(e)\rangle = \langle a(e), a(e)\rangle$$

and thus $a(e) = 0$. Because a is Γ-linear, we also obtain $a(g \cdot e) = g \cdot a(e) = 0$. Continuity of a therefore shows that $a = 0$.

Ad 3. This is clear from the definition of positivity and the trace. $\qquad\square$

Example 1.1.13 (some von Neumann traces).

- If Γ is a finite group, then $N\Gamma = \mathbb{C}\Gamma$. The von Neumann trace is

$$\mathrm{tr}_\Gamma \colon N\Gamma = \mathbb{C}\Gamma \longrightarrow \mathbb{C}$$
$$\sum_{g \in \Gamma} a_g \cdot g \longmapsto a_e.$$

- If $\Gamma = \mathbb{Z} = \langle t \mid \rangle$, then we obtain [105, Example 1.4]: The group von Neumann algebra $N\Gamma$ is canonically isomorphic to $L^\infty([-\pi,\pi],\mathbb{C})$ (as can be seen via the Fourier transform) and the action on $\ell^2\Gamma \cong L^2([-\pi,\pi],\mathbb{C})$ is given by pointwise multiplication; under this isomorphism, the trace tr_Γ on $N\Gamma$ corresponds to the integration map

$$L^\infty([-\pi,\pi],\mathbb{C}) \longrightarrow \mathbb{C}$$
$$f \longmapsto \frac{1}{2\pi} \cdot \int_{[-\pi,\pi]} f \, d\lambda.$$

In view of the previous example (and Theorem A.1.3), the abstract theory of von Neumann algebras is also sometimes referred to as *non-commutative measure theory*.

Remark 1.1.14 (extension of the trace to matrices and morphisms). As in linear algebra, we can extend the trace from the group von Neumann algebra to matrices: Let Γ be a countable group and let $n \in \mathbb{N}$. Then we define the trace

$$\mathrm{tr}_\Gamma \colon M_{n \times n}(N\Gamma) \longrightarrow \mathbb{C}$$

$$A \longmapsto \sum_{j=1}^{n} \mathrm{tr}_\Gamma A_{jj}.$$

This trace also satisfies the trace property, is faithful, and positive (check!).

Moreover, every bounded (left) Γ-equivariant map $(\ell^2\Gamma)^n \longrightarrow (\ell^2\Gamma)^n$ is represented by a matrix in $M_{n \times n}(N\Gamma)$ (by the last characterisation in Remark 1.1.11). Therefore, every bounded Γ-equivariant map $(\ell^2\Gamma)^n \longrightarrow (\ell^2\Gamma)^n$ has a trace.

1.2 The von Neumann dimension

We can now define the von Neumann dimension of Hilbert modules via the trace of projections:

Proposition and Definition 1.2.1 (von Neumann dimension). *Let Γ be a countable group and let V be a Hilbert Γ-module. Then the* von Neumann Γ-dimension *of V is defined as*

$$\dim_{N\Gamma} V := \mathrm{tr}_\Gamma p,$$

where $i\colon V \longrightarrow (\ell^2\Gamma)^n$ (for some $n \in \mathbb{N}$) is an isometric Γ-embedding and $p\colon (\ell^2\Gamma)^n \longrightarrow (\ell^2\Gamma)^n$ is the orthogonal Γ-projection onto $i(V)$. This is well-defined (i.e., independent of the chosen embedding into a finitely generated free $\ell^2\Gamma$-module) and $\dim_{N\Gamma} V \in \mathbb{R}_{\geq 0}$.

Proof. As a first step, we note that $i(V)$ is a *closed* subspace of $(\ell^2\Gamma)^n$ (because V is complete and i is isometric). Hence, there indeed exists an orthogonal projection $p\colon (\ell^2\Gamma)^n \longrightarrow \mathrm{im}\, i$.

The trace is independent of the embedding: Let $j\colon V \longrightarrow (\ell^2\Gamma)^m$ also be an isometric Γ-embedding and let $q\colon (\ell^2\Gamma)^m \longrightarrow \mathrm{im}\, j$ be the orthogonal projection. Then we define a partial isometry $u\colon (\ell^2\Gamma)^n \longrightarrow (\ell^2\Gamma)^m$ by taking $j \circ i^{-1}$ on $\mathrm{im}\, i$ and taking 0 on $(\mathrm{im}\, i)^\perp$. By construction $j = u \circ i$. Taking adjoints shows that $q = p \circ u^*$ and hence

$$\mathrm{tr}_\Gamma q = \mathrm{tr}_\Gamma(j \circ q) = \mathrm{tr}_\Gamma(u \circ i \circ q) = \mathrm{tr}_\Gamma(u \circ i \circ p \circ u^*)$$
$$= \mathrm{tr}_\Gamma(i \circ p \circ u^* \circ u) \qquad\qquad \text{(trace property)}$$
$$= \mathrm{tr}_\Gamma(i \circ p \circ p) = \mathrm{tr}_\Gamma(i \circ p) = \mathrm{tr}_\Gamma p.$$

The von Neumann dimension is non-negative: Let $P \in M_{n \times n}(N\Gamma)$ be the matrix representing p. Because p (as an orthogonal projection) is a positive operator (check!), all the diagonal entries $P_{jj} \in N\Gamma$ of P are also positive operators (check!). Therefore, positivity of the von Neumann trace (Theorem 1.1.12) shows that $\dim_{N\Gamma} V = \mathrm{tr}_\Gamma p = \sum_{j=1}^{n} \mathrm{tr}_\Gamma P_{jj} \geq 0$. $\qquad\square$

Example 1.2.2 (von Neumann dimension).

- Let Γ be a finite group and let V be a Hilbert Γ-module. Then

$$\dim_{N\Gamma} V = \frac{1}{|\Gamma|} \cdot \dim_{\mathbb{C}} V,$$

as can be seen from a direct computation (check!) or by applying the restriction formula (Theorem 1.2.3).

- Let $\Gamma = \mathbb{Z} = \langle t \,|\, \rangle$. We will use the description of $\ell^2\Gamma$ and $N\Gamma$ from Example 1.1.13. Let $A \subset [-\pi, \pi]$ be a measurable set. Then $V := \{f \cdot \chi_A \mid f \in L^2([-\pi, \pi], \mathbb{C})\}$ is a Hilbert Γ-module and the 1×1-matrix $(\chi_A) \in M_{1 \times 1}(N\Gamma)$ describes the orthogonal projection onto V. Hence,

$$\dim_{N\Gamma} V = \operatorname{tr}_\Gamma \chi_A = \frac{1}{2\pi} \cdot \int_{-\pi}^{\pi} \chi_A \, d\lambda = \frac{1}{2\pi} \cdot \lambda(A)$$

and thus every number in $[0, 1]$ occurs as the von Neumann dimension of a Hilbert $N\Gamma$-module (!).

Theorem 1.2.3 (basic properties of the von Neumann dimension). *Let Γ be a countable group.*

1. Normalisation. *We have $\dim_{N\Gamma} \ell^2\Gamma = 1$.*

2. Faithfulness. *For every Hilbert Γ-module V, we have $\dim_{N\Gamma} V = 0$ if and only if $V \cong_\Gamma 0$.*

3. Weak isomorphism invariance. *If $f \colon V \longrightarrow W$ is a morphism of Hilbert Γ-modules with $\ker f = 0$ and $\overline{\operatorname{im} f} = W$, then $\dim_\Gamma V = \dim_\Gamma W$.*

4. Additivity. *Let $0 \longrightarrow V' \overset{i}{\longrightarrow} V \overset{\pi}{\longrightarrow} V'' \longrightarrow 0$ be a weakly exact sequence of Hilbert Γ-modules (i.e., i is injective, $\overline{\operatorname{im} i} = \ker \pi$ and $\overline{\operatorname{im} \pi} = V''$). Then*

$$\dim_{N\Gamma} V = \dim_{N\Gamma} V' + \dim_{N\Gamma} V''.$$

5. Multiplicativity. *Let Λ be a countable group, let V be a Hilbert Γ-module, and let W be a Hilbert Λ-module. Then the completed tensor product $V \overline{\otimes}_{\mathbb{C}} W$ is a Hilbert $\Gamma \times \Lambda$-module and*

$$\dim_{N(\Gamma \times \Lambda)}(V \overline{\otimes}_{\mathbb{C}} W) = \dim_{N\Gamma} V \cdot \dim_{N\Lambda} W.$$

6. Restriction. *Let V be a Hilbert Γ-module and let $\Lambda \subset \Gamma$ be a subgroup of finite index. Then*

$$\dim_{N\Lambda} \operatorname{Res}_\Lambda^\Gamma V = [\Gamma : \Lambda] \cdot \dim_{N\Gamma} V.$$

Proof. Ad 1. This is clear from the definition (we can take $\mathrm{id}_{\ell^2\Gamma}$ as embedding and projection).

Ad 2. In view of faithfulness of the von Neumann trace (Theorem 1.1.12), it follows that the von Neumann trace of a projection is 0 if and only if the projection is 0 (check!).

Ad 3. This is a consequence of polar decomposition: Let $f = u \circ p$ be the polar decomposition of f into a partial isometry u and a positive operator p with $\ker u = \ker p$. We now show that u is a Γ-isometry between V and W: As f is injective, we have $\ker u = \ker p = 0$. Moreover, as a partial isometry, u has closed image and so $\operatorname{im} u = \overline{\operatorname{im} u} = \overline{\operatorname{im} f} = W$. Hence, u is an isometry. Moreover, the uniqueness of the polar decomposition shows that u is Γ-equivariant. Therefore, $\dim_{N\Gamma} V = \dim_{N\Gamma} W$.

Ad 4. The von Neumann dimension is additive with respect to direct sums (check!). Moreover,

$$V \longrightarrow \overline{\operatorname{im} i} \oplus V''$$
$$x \longmapsto (p(x), \pi(x))$$

is a weak isomorphism of Hilbert Γ-modules (check!), where $p \colon V \longrightarrow \overline{\operatorname{im} i}$ denotes the orthogonal projection. Therefore, weak isomorphism invariance of $\dim_\Gamma$ shows that

$$\dim_\Gamma V = \dim_\Gamma(\overline{\operatorname{im} i} \oplus V'') = \dim_\Gamma \overline{\operatorname{im} i} + \dim_\Gamma V'' = \dim_\Gamma V' + \dim_\Gamma V''.$$

Ad 5. The key observation is that $\ell^2(\Gamma \times \Lambda)$ is isomorphic (as a Hilbert $\Gamma \times \Lambda$-module) to $\ell^2\Gamma \,\overline{\otimes}_{\mathbb{C}}\, \ell^2\Lambda$ [105, Theorem 1.12].

Ad 6. This is Exercise 1.E.4.   □

Moreover, the von Neumann dimension also satisfies inner and outer regularity [105, Theorem 1.12].

Outlook 1.2.4 (the extended von Neumann dimension). The above hands-on construction of the von Neumann dimension is convenient for simple computations. However, this approach only works for the category of Hilbert modules, which in general is only additive but not Abelian. Several extensions of the von Neumann dimension are available, e.g., by Cheeger and Gromov [36], Farber [51], and Lück [104, 105]. We will briefly outline Lück's algebraic version:

Let Γ be a countable group. Then the category of Hilbert Γ-modules canonically embeds into the category of $N\Gamma$-modules (Exercise 1.E.2); moreover, this construction can be refined to a $\mathbb{C}$-linear equivalence F from the category of Hilbert Γ-modules to the category of finitely generated projective $N\Gamma$-modules that satisfies $F(\ell^2\Gamma) = N\Gamma$ and preserves (weak) exactness [104, Theorem 1.8].

If P is a finitely generated projective $N\Gamma$-module, then we set

$$\text{pdim}_{N\Gamma} P := \text{tr}_{N\Gamma} p := \text{tr}_{\Gamma} A \in \mathbb{R}_{\geq 0},$$

where $p \colon (N\Gamma)^n \longrightarrow (N\Gamma)^n$ is a projection with $P \cong_{N\Gamma} \text{im}\, p$ and associated matrix $A \in M_{n \times n}(N\Gamma)$; this definition is independent of the chosen projection p [105, p. 238f].

For a general $N\Gamma$-module V one then defines

$$\dim_{N\Gamma} V := \sup\{\text{pdim}_{N\Gamma} P \mid P \text{ is a finitely generated projective}$$
$$N\Gamma\text{-submodule of } V\} \in \mathbb{R}_{\geq 0} \cup \{\infty\}.$$

It turns out that this definition provides a well-behaved notion of dimension for $N\Gamma$-modules that coincides via F with the von Neumann dimension of Hilbert Γ-modules [104, Theorem 0.6]. One of the key ingredients is the observation that the ring $N\Gamma$ is semi-hereditary (i.e., that every finitely generated submodule of a projective $N\Gamma$-module is projective).

Furthermore, the $N\Gamma$-modules of dimension 0 form a Serre subcategory of the category of $N\Gamma$-modules; this allows us to efficiently use standard tools from homological algebra when working with von Neumann dimensions [123].

In fact, the same construction works for every finite von Neumann algebra, not only for the group von Neumann algebra $N\Gamma$ [104, 123]. In Chapter 4, we will use such an extended von Neumann dimension in the context of equivalence relations.

Outlook 1.2.5 (Atiyah conjecture). The Atiyah question/conjecture comes in many flavours (originally formulated in terms of closed Riemannian manifolds). One version is:

Let Γ be a torsion-free countable group, let $n \in \mathbb{N}$, and let $A \in M_{n \times n}(\mathbb{C}\Gamma)$. Then $\dim_{N\Gamma} \ker M_A \in \mathbb{Z}$ (?!)

This version of the Atiyah conjecture is known to hold for many classes of groups (and no counterexample is known so far); however, more general versions of the Atiyah conjecture are known to be false [12, 72]. One interesting aspect of the Atiyah conjecture is that it implies the Kaplansky zero-divisor conjecture (Exercise 1.E.5).

1.E Exercises

Exercise 1.E.1 (the "trivial" Hilbert module). For which countable groups Γ is $\mathbb{C}$ (with the trivial Γ-action) a Hilbert Γ-module? Which von Neumann dimension does it have?

Exercise 1.E.2 (Hilbert modules as modules over the von Neumann algebra). Let Γ be a countable group and let V be a Hilbert Γ-module. Show that the left Γ-action on V extends to a left $N\Gamma$-action on V. How can this construction be turned into a functor?

Hints. This fact is the reason why Hilbert Γ-modules are often called *Hilbert $N\Gamma$-modules*.

Exercise 1.E.3 (kernels and cokernels). Let Γ be a countable group, let V and W be Hilbert Γ-modules, and let $\varphi \colon V \longrightarrow W$ be a morphism of Hilbert Γ-modules.

1. Show that $\ker \varphi$ (with the induced inner product and Γ-action) is a Hilbert Γ-module

2. Show that $W/\overline{\operatorname{im} \varphi}$ (with the induced inner product and Γ-action) is a Hilbert Γ-module.

   Hints. Orthogonal complement!

Exercise 1.E.4 (restriction formula for the von Neumann dimension [105, Theorem 1.12(6)]). Let Γ be a countable group, let V be a Hilbert Γ-module, and let $\Lambda \subset \Gamma$ be a finite index subgroup. Show that

$$\dim_{N\Lambda} \operatorname{Res}_{\Lambda}^{\Gamma} V = [\Gamma : \Lambda] \cdot \dim_{N\Gamma} V.$$

Exercise 1.E.5 (Atiyah $\Longrightarrow$ Kaplansky [105, Lemma 10.15]). Let Γ be a countable torsion-free group that satisfies the Atiyah conjecture (Outlook 1.2.5). Show that $\mathbb{C}\Gamma$ is a domain.

2

L^2-Betti numbers

We will now use the von Neumann dimension to define L^2-Betti numbers of groups and spaces. We will study basic properties of L^2-Betti numbers, and we will compute some simple examples. Furthermore, we will briefly outline generalisations of L^2-Betti numbers beyond our elementary approach.

Overview of this chapter.

2.1	An elementary definition of L^2-Betti numbers	18
2.2	Basic computations	20
2.3	Variations and extensions	24
2.E	Exercises	26

Running example. the additive group $\mathbb{Z}$, finite (index sub)groups, free groups

© The Author(s), under exclusive license to Springer Nature Switzerland AG 2020
C. Löh, *Ergodic Theoretic Methods in Group Homology*,
SpringerBriefs in Mathematics, https://doi.org/10.1007/978-3-030-44220-0_3

2.1 An elementary definition of L^2-Betti numbers

L^2-Betti numbers are an equivariant version of ordinary Betti numbers. For simplicity, we will only consider L^2-Betti numbers of free Γ-CW-complexes of finite type.

2.1.1 Finite type

Definition 2.1.1 (equivariant CW-complex). Let Γ be a group.

- A *free Γ-CW-complex* is a CW-complex X together with a free Γ-action such that:
    - the Γ-action permutes the open cells of X and
    - if e is an open cell of X and $g \in \Gamma$ is non-trivial, then $g \cdot e \cap e \neq \emptyset$.
- A *morphism of Γ-CW-complexes* is a Γ-equivariant cellular map.

Definition 2.1.2 (finite type).

- A CW-complex is of *finite type* if for each $n \in \mathbb{N}$, there are only finitely many open n-cells.

- Let Γ be a group. A (free) Γ-CW-complex is of *finite type* if for each dimension $n \in \mathbb{N}$, there are only finitely many Γ-orbits of open n-cells.

- A group Γ is of *finite type* if it admits a classifying space of finite type (equivalently, a classifying space whose universal covering with the induced free Γ-CW-structure is a Γ-CW-complex of finite type).

Remark 2.1.3 (more on groups of finite type). Let Γ be a group.

- If Γ is of finite type, then Γ is finitely presented (as the fundamental group of a CW-complex of finite type); in particular, Γ is countable.

- If Γ is finitely presented, then Γ is of finite type if and only if $\mathbb{C}$ (with the trivial Γ-action) admits a projective resolution over $\mathbb{C}\Gamma$ that is finitely generated in each degree [30, Chapter VIII].

Example 2.1.4 (groups of finite type).

- The group $\mathbb{Z}$ is of finite type: We can take the circle (with our favourite CW-structure) as a classifying space.

- If Γ and Λ are of finite type, then so is $\Gamma \times \Lambda$ (we can take the product of finite type models as a model for the classifying space of $\Gamma \times \Lambda$).

- If Γ and Λ are of finite type, then so is $\Gamma * \Lambda$ (we can take the wedge of finite type models as a model for the classifying space of $\Gamma * \Lambda$).

- In particular, free Abelian groups of finite rank and free groups of finite rank are of finite type.

- Let $g \in \mathbb{N}_{\geq 2}$ and let Σ_g be "the" oriented closed connected surface of genus g. Then $\pi_1(\Sigma_g)$ is of finite type (because Σ_g is a (finite) model for the classifying space of Σ_g).

- More generally: If M is an oriented closed connected hyperbolic manifold, then $\pi_1(M)$ has finite type (because M is a (finite) model for the classifying space of $\pi_1(M)$).

- If $n \in \mathbb{N}_{\geq 2}$, then $\mathbb{Z}/n$ is of finite type (check!), but there is *no* finite model for the classifying space of $\mathbb{Z}/n$ [30, Corollary VIII.2.5].

- More generally: All finite groups are of finite type (for instance, one can use the simplicial Γ-resolution as a blueprint to construct a contractible free Γ-CW-complex and then take its quotient).

- There exist finitely presented groups that are *not* of finite type [20].

2.1.2 L^2-Betti numbers of spaces

Definition 2.1.5 (L^2-Betti numbers of spaces). Let Γ be a countable group and let X be a free Γ-CW-complex of finite type.

- The *cellular L^2-chain complex of X* is the twisted chain complex

$$C_*^{(2)}(\Gamma \curvearrowright X) := \ell^2\Gamma \otimes_{\mathbb{C}\Gamma} C_*(X).$$

Here, $C_*(X)$ denotes the cellular chain complex of X (with $\mathbb{C}$-coefficients) with the induced Γ-action and $\ell^2\Gamma$ carries the left $\mathbb{C}\Gamma$-module structure given by right translation on Γ.

- Let $n \in \mathbb{N}$. The *(reduced) L^2-homology of X in degree n* is defined by

$$H_n^{(2)}(\Gamma \curvearrowright X) := \ker \partial_n^{(2)} \,/\, \overline{\operatorname{im} \partial_{n+1}^{(2)}},$$

where $\partial_*^{(2)}$ denotes the boundary operator on $C_*^{(2)}(\Gamma \curvearrowright X)$.

- The *n-th L^2-Betti number of X* is defined by

$$b_n^{(2)}(\Gamma \curvearrowright X) := \dim_{N\Gamma} H_n^{(2)}(\Gamma \curvearrowright X),$$

where $\dim_{N\Gamma}$ is the von Neumann dimension (Definition 1.2.1). It should be noted that $H_n^{(2)}(\Gamma \curvearrowright X)$ is indeed a Hilbert Γ-module (this follows from Exercise 1.E.3).

Notation 2.1.6. Moreover, we use the following abbreviation: If X is a CW-complex of finite type with fundamental group Γ and universal covering $\tilde{X}$ (with the induced free Γ-CW-complex structure), then we write

$$b_n^{(2)}(X) := b_n^{(2)}(\Gamma \curvearrowright \tilde{X}).$$

It should be noted that in the literature the notation $b_n^{(2)}(\tilde{X})$ can also be found as an abbreviation for $b_n^{(2)}(\Gamma \curvearrowright \tilde{X})$. However, we prefer the notation $b_n^{(2)}(X)$ as it is less ambiguous (what is $b_n^{(2)}(\mathbb{H}^2)$?!).

Remark 2.1.7 (homotopy invariance). Let Γ be a countable group, let X and Y be free Γ-CW-complexes, and let $n \in \mathbb{N}$. If $f\colon X \longrightarrow Y$ is a (cellular) Γ-homotopy equivalence, then

$$b_n^{(2)}(\Gamma \curvearrowright X) = b_n^{(2)}(\Gamma \curvearrowright Y),$$

because f induces a $\mathbb{C}\Gamma$-chain homotopy equivalence $C_*(X) \simeq_{\mathbb{C}\Gamma} C_*(Y)$ and thus a chain homotopy equivalence $C_*^{(2)}(\Gamma \curvearrowright X) \longrightarrow C_*^{(2)}(\Gamma \curvearrowright Y)$ in the category of chain complexes of Hilbert Γ-modules.

2.1.3 L^2-Betti numbers of groups

Let Γ be a group and let X and Y be models for the classifying space of Γ. Then the universal coverings $\tilde{X}$ and $\tilde{Y}$ with the induced Γ-CW-structures are (cellularly) Γ-homotopy equivalent. Therefore, $b_n^{(2)}(X) = b_n^{(2)}(Y)$ for all $n \in \mathbb{N}$ (Remark 2.1.7). Hence, the following notion is well-defined:

Definition 2.1.8 (L^2-Betti numbers of groups). Let Γ be a group of finite type and let $n \in \mathbb{N}$. Then the *n-th L^2-Betti number of* Γ is defined by

$$b_n^{(2)}(\Gamma) := b_n^{(2)}(X),$$

where X is a model for the classifying space of Γ of finite type.

Similarly, we could also define/compute L^2-Betti numbers of groups by tensoring finite type projective resolutions of $\mathbb{C}$ over $\mathbb{C}\Gamma$ with $\ell^2\Gamma$ (check!).

2.2 Basic computations

For simplicity, in the following, we will focus on L^2-Betti numbers of groups. Similar statements hold for L^2-Betti numbers of spaces [105, Theorem 1.35].

2.2.1 Basic properties

Proposition 2.2.1 (degree 0). *Let Γ be a group of finite type.*

1. *If Γ is finite, then $b_0^{(2)}(\Gamma) = 1/|\Gamma|$.*

2. *If Γ is infinite, then $b_0^{(2)}(\Gamma) = 0$.*

Both cases can conveniently be summarised in the formula

$$b_0^{(2)}(\Gamma) = \frac{1}{|\Gamma|}.$$

Proof. Classical group homology tells us that $b_0^{(2)}(\Gamma) = \dim_{N\Gamma} V$, where

$$V = \ell^2\Gamma \,/\, \overline{\mathrm{Span}_{\mathbb{C}}\{x - g \cdot x \mid x \in \ell^2\Gamma,\ g \in \Gamma\}}.$$

If Γ is *finite*, then $\ell^2\Gamma = \mathbb{C}\Gamma$, whence $V \cong_\Gamma \mathbb{C}$ (with the trivial Γ-action). Therefore (Exercise 1.E.1),

$$b_0^{(2)}(\Gamma) = \dim_{N\Gamma} \mathbb{C} = \frac{1}{|\Gamma|}.$$

If Γ is *infinite*, then it suffices to show that $V \cong_\Gamma 0$: To this end, we only need to show that every bounded $\mathbb{C}$-linear functional $V \longrightarrow \mathbb{C}$ is the zero functional. Equivalently, we need to show that every Γ-invariant bounded $\mathbb{C}$-linear functional $f\colon \ell^2\Gamma \longrightarrow \mathbb{C}$ satisfies $f|_\Gamma = 0$ (check!). As Γ is infinite (and countable), we can enumerate Γ as $(g_n)_{n\in\mathbb{N}}$. The element $x := \sum_{n\in\mathbb{N}} 1/n \cdot g_n$ lies in $\ell^2\Gamma$ and the computation

$$f(x) = \sum_{n\in\mathbb{N}} \frac{1}{n} \cdot f(g_n) \qquad \text{(continuity and linearity of } f\text{)}$$

$$= \sum_{n\in\mathbb{N}} \frac{1}{n} \cdot f(e) \qquad \text{(} \Gamma\text{-invariance of } f\text{)}$$

shows that $f(e) = 0$ (otherwise the series would *not* converge). Hence $f(g) = f(e) = 0$ for all $g \in \Gamma$ (by Γ-invariance), as desired. $\qquad\square$

Theorem 2.2.2 (inheritance properties of L^2-Betti numbers). *Let Γ be a group of finite type and let $n \in \mathbb{N}$.*

1. *Dimension. If Γ admits a finite model of the classifying space of dimension less than n, then $b_n^{(2)}(\Gamma) = 0$.*

2. *Restriction. If $\Lambda \subset \Gamma$ is a subgroup of finite index, then Λ is also of finite type and*

$$b_n^{(2)}(\Lambda) = [\Gamma : \Lambda] \cdot b_n^{(2)}(\Gamma).$$

3. Künneth formula. *If Λ is a group of finite type, then*

$$b_n^{(2)}(\Gamma \times \Lambda) = \sum_{j=0}^{n} b_j^{(2)}(\Gamma) \cdot b_{n-j}^{(2)}(\Lambda).$$

4. Additivity. *If Λ is a group of finite type, then*

$$b_1^{(2)}(\Gamma * \Lambda) = b_1^{(2)}(\Gamma) + b_1^{(2)}(\Lambda) + 1 - \left(b_0^{(2)}(\Gamma) + b_0^{(2)}(\Lambda)\right)$$

and, if $n > 1$, then

$$b_n^{(2)}(\Gamma * \Lambda) = b_n^{(2)}(\Gamma) + b_n^{(2)}(\Lambda).$$

5. Poincaré duality. *If Γ admits a classifying space that is an oriented closed connected d-manifold, then*

$$b_n^{(2)}(\Gamma) = b_{d-n}^{(2)}(\Gamma).$$

Proof. Ad 1. This is clear from the definition.

Ad 2. If X is a finite type model for the classifying space of Γ, then the covering space Y associated with the subgroup $\Lambda \subset \Gamma$ is a model for the classifying space of Λ; moreover, Y is of finite type because the covering degree is $[\Gamma : \Lambda]$, which is finite. Algebraically, on the cellular chain complex of the universal covering space $\widetilde{X} = \widetilde{Y}$ (and whence also on its reduced cohomology with ℓ^2-coefficients), this corresponds to applying the restriction functor $\mathrm{Res}_\Lambda^\Gamma$. Then, we only need to apply the restriction formula for the von Neumann dimension (Theorem 1.2.3).

Ad 3. Let X and Y be finite type models for the classifying space of Γ and Λ, respectively. Then $X \times Y$ is a model for the classifying space of $\Gamma \times \Lambda$, which is of finite type. One can now use a Künneth argument and the multiplicativity of the von Neumann dimension (Theorem 1.2.3) to prove the claim [105, Theorem 1.35].

Ad 4. Let X and Y be finite type models for the classifying space of Γ and Λ, respectively. Then the wedge $X \vee Y$ is a model for the classifying space of $\Gamma * \Lambda$, which is of finite type. One can now use a (cellular) Mayer–Vietoris argument to prove the additivity formula [105, Theorem 1.35].

Ad 5. The main ingredients are twisted Poincaré duality (applied to the coefficients $\ell^2\Gamma$) and the fact that the L^2-Betti numbers can also be computed in terms of reduced cohomology [105, Theorem 1.35]. $\square$

Proposition 2.2.3 (Euler characteristic). *Let Γ be a group that admits a finite classifying space. Then*

$$\chi(\Gamma) = \sum_{n \in \mathbb{N}} (-1)^n \cdot b_n^{(2)}(\Gamma).$$

Proof. This follows (as in the classical case) from the additivity of the von Neumann dimension (Exercise 2.E.2). □

2.2.2 First examples

Example 2.2.4 (finite groups). Let Γ be a finite group. Then Γ is of finite type (Example 2.1.4) and, for all $n \in \mathbb{N}$, we have

$$b_n^{(2)}(\Gamma) = \frac{1}{|\Gamma|} \cdot \dim_{\mathbb{C}} H_n(\Gamma; \mathbb{C}\Gamma) = \begin{cases} \frac{1}{|\Gamma|} & \text{if } n = 0 \\ 0 & \text{if } n > 0. \end{cases}$$

Example 2.2.5 (the additive group $\mathbb{Z}$). There are many ways to see that the L^2-Betti numbers of the additive group $\mathbb{Z}$ are all equal to 0. For instance: Let $n \in \mathbb{N}$. For $k \in \mathbb{N}_{>1}$, we consider the subgroup $k \cdot \mathbb{Z} \subset \mathbb{Z}$ of index k. Then the restriction formula (Theorem 2.2.2) shows that

$$\begin{aligned} b_n^{(2)}(\mathbb{Z}) &= b_n^{(2)}(k \cdot \mathbb{Z}) && (\text{because } k \cdot \mathbb{Z} \cong_{\mathsf{Group}} \mathbb{Z}) \\ &= k \cdot b_n^{(2)}(\mathbb{Z}) && (\text{restriction formula}) \end{aligned}$$

and so $b_n^{(2)}(\mathbb{Z}) = 0$.

Example 2.2.6 (free groups). Let $r \in \mathbb{N}_{\geq 1}$ and let F_r be "the" free group of rank r. Then $X_r := \bigvee^r S^1$ is a model of the classifying space of F_r.

- Because F_r is infinite, we have $b_0^{(2)}(F_r) = 0$ (Proposition 2.2.1).

- Because $\dim X_r = 1$, we have $b_n^{(2)}(F_r) = 0$ for all $n \in \mathbb{N}_{\geq 2}$.

- It thus remains to compute $b_1^{(2)}(F_r)$. Because the Euler characteristic can be calculated via L^2-Betti numbers (Proposition 2.2.3), we obtain

$$b_1^{(2)}(F_r) = -\chi(F_r) + b_0^{(2)}(F_r) = -\chi(X_r) + 0 = r - 1.$$

Example 2.2.7 (surface groups). Let $g \in \mathbb{N}_{\geq 1}$ and let $\Gamma_g := \pi_1(\Sigma_g)$, where Σ_g is "the" oriented closed connected surface of genus g. Then Σ_g is a model of the classifying space of Σ_g.

- Because Γ_g is infinite, we have $b_0^{(2)}(\Gamma_g) = 0$ (Proposition 2.2.1).

- Because $\dim \Sigma_g = 2$, we obtain from Poincaré duality (Theorem 2.2.2) that $b_2^{(2)}(\Gamma_g) = b_0^{(2)}(\Gamma_g) = 0$ and that $b_n^{(2)}(\Gamma_g) = 0$ for all $n \in \mathbb{N}_{\geq 3}$.

- It thus remains to compute $b_1^{(2)}(\Sigma_g)$. Because the Euler characteristic can be calculated via L^2-Betti numbers (Proposition 2.2.3), we obtain

$$b_1^{(2)}(\Gamma_g) = -\chi(\Gamma_g) + b_0^{(2)}(\Gamma_g) + b_2^{(2)}(\Gamma_g) = -\chi(\Sigma_g) + 0 = 2 \cdot (g-1).$$

Outlook 2.2.8 (hyperbolic manifolds). More generally: Let Γ be the fundamental group of an oriented closed connected hyperbolic manifold M of dimension d.

- If d is odd, then $b_n^{(2)}(\Gamma) = 0$ for all $n \in \mathbb{N}$.

- If d is even, then $b_n^{(2)}(\Gamma) = 0$ for all $n \in \mathbb{N} \setminus \{d/2\}$. Moreover,

$$b_{d/2}^{(2)}(\Gamma) \neq 0.$$

The proof is based on the fact that L^2-Betti numbers can be computed in terms of spaces of L^2-harmonic forms [46][105, Chapter 1.4] and the explicit computation of L^2-harmonic forms of hyperbolic manifolds [47][105, Theorem 1.62]. (A cellular version of harmonic forms is discussed in Exercise 3.E.2.)

Outlook 2.2.9 (Singer conjecture). The Singer conjecture predicts that L^2-Betti numbers of closed aspherical manifolds are concentrated in the middle dimension:

> Let M be an oriented closed connected aspherical manifold of dimension d. Then
> $$\forall_{n \in \mathbb{N} \setminus \{d/2\}} \quad b_n^{(2)}(\pi_1(M)) = 0 \quad (?!)$$

No counterexample is known. However, the analogue of the Singer conjecture for rationally aspherical manifolds is false [13] and not much is known about L^2-Betti numbers of "exotic" closed aspherical manifolds [41].

2.3 Variations and extensions

- **Analytic definition.** Originally, Atiyah defined L^2-Betti numbers (of closed smooth manifolds) in terms of the heat kernel on the universal covering [11]. Dodziuk proved that these analytic L^2-Betti numbers admit a combinatorial description (in terms of ℓ^2-chain complexes of simplicial/cellular complexes of finite type) [46].

- **Singular definition.** Cheeger and Gromov [36], Farber [51], Lück [104, 105] extended the definition of the von Neumann dimension to all modules over the von Neumann algebra (Outlook 1.2.4); in particular, this allows for a definition of L^2-Betti numbers of spaces in terms of singular homology with twisted coefficients in $\ell^2\pi_1$ or $N\pi_1$.

- **Extension to equivalence relations.** Gaboriau extended the definition of L^2-Betti numbers of groups to standard equivalence relations [64]. We will return to this point of view in Chapter 4.

- **Extension to topological groups.** Petersen gave a definition of L^2-Betti numbers of locally compact, second countable, unimodular groups [119].

- **Version for von Neumann algebras.** In a slightly different direction, Connes and Shlyakhtenko introduced a notion of L^2-Betti numbers for tracial von Neumann algebras [40]. However, it is unknown to what extent these L^2-Betti numbers of group von Neumann algebras coincide with the L^2-Betti numbers of groups (which would be helpful in the context of the free group factor isomorphism problem).

It should be noted that Popa and Vaes showed that Thom's continuous version of the corresponding L^2-cohomology for tracial von Neumann algebras is always trivial in degree 1 [121].

2.E Exercises

Exercise 2.E.1 (products).

1. Let Γ and Λ be infinite groups of finite type. Show that $b_1^{(2)}(\Gamma \times \Lambda) = 0$.

2. Conclude: If Γ is a group of finite type with $b_1^{(2)}(\Gamma) \neq 0$, then Γ does *not* contain a finite index subgroup that is a product of infinite groups.

Exercise 2.E.2 (Euler characteristic). Let Γ be a group that admits a finite classifying space. Show that

$$\chi(\Gamma) = \sum_{n \in \mathbb{N}} (-1)^n \cdot b_n^{(2)}(\Gamma).$$

Exercise 2.E.3 (an explicit ℓ^2-cycle [84, Figure 1.1]). Give an explicit example of a non-zero 1-cycle in $C_*^{(2)}(F_2 \curvearrowright T)$, where T is "the" regular 4-valent tree.

Exercise 2.E.4 (QI?! [135]). Let Γ be a group of finite type and let $r \in \mathbb{N}_{\geq 2}$.

1. Compute all L^2-Betti numbers of $F_r * \Gamma$ in terms of r and the L^2-Betti numbers of Γ.

2. Let $k \in \mathbb{N}_{\geq 2}$. Conclude that the quotient $b_1^{(2)}/b_k^{(2)}$ is *not* a quasi-isometry invariant.

3. Show that the sign of the Euler characteristic is *not* a quasi-isometry invariant.

4. Use these results to prove that there exist groups of finite type that are quasi-isometric but not commensurable.

Hints. If $s \in \mathbb{N}_{\geq 2}$, then it is known that F_r and F_s are bilipschitz equivalent and thus that $F_r * \Gamma$ and $F_s * \Gamma$ are quasi-isometric [115, 135].

Exercise 2.E.5 (deficiency [102]).

1. Let Γ be a group of finite type and let $\langle S \mid R \rangle$ be a finite presentation of Γ. Show that

$$|S| - |R| \leq 1 - b_0^{(2)}(\Gamma) + b_1^{(2)}(\Gamma) - b_2^{(2)}(\Gamma).$$

Taking the maximum of all these differences thus shows that the *deficiency* def(Γ) *of* Γ is bounded from above by the right-hand side.

2. Let $\Gamma \subset \text{Isom}^+(\mathbb{H}^4)$ be a torsion-free uniform lattice. Show that

$$\text{def}(\Gamma) \leq 1 - \chi(\Gamma) = 1 - \frac{3}{4 \cdot \pi^2} \cdot \text{vol}(\Gamma \backslash \mathbb{H}^4).$$

3

The residually finite view: Approximation

The L^2-Betti numbers are related to classical Betti numbers through approximation by the normalised Betti numbers of finite index subgroups/finite coverings.

We explain the (spectral) proof of this approximation theorem and briefly discuss the relation with other (homological) gradient invariants.

This residually finite view will be complemented by the dynamical view in Chapter 4 and the approximation theorems for lattices in Chapter 5.

Overview of this chapter.

3.1	The approximation theorem	28
3.2	Proof of the approximation theorem	28
3.3	Homological gradient invariants	32
3.E	Exercises	35

Running example. free Abelian groups, free groups

3.1 The approximation theorem

In the residually finite view, one approximates groups by finite quotients/finite index subgroups and spaces by finite coverings.

Definition 3.1.1 (residual chain, residually finite group). Let Γ be a finitely generated group.

- A *residual chain* for Γ is a sequence $(\Gamma_n)_{n\in\mathbb{N}}$ of finite index normal subgroups of Γ with $\Gamma_0 \supset \Gamma_1 \supset \Gamma_2 \supset \dots$ and $\bigcap_{n\in\mathbb{N}} \Gamma_n = \{e\}$.

- The group Γ is *residually finite* if it admits a residual chain.

Example 3.1.2 (residually finite groups).

- All finitely generated linear groups are residually finite [107, 113]. In particular: Fundamental groups of closed hyperbolic manifolds are residually finite. In contrast, it is unknown whether all finitely generated (Gromov-)hyperbolic groups are residually finite.

- There exist finitely presented groups that are not residually finite (e.g., each finitely presented infinite simple group will do).

Theorem 3.1.3 (Lück's approximation theorem [103]). *Let X be a connected CW-complex of finite type with residually finite fundamental group Γ, let $(\Gamma_n)_{n\in\mathbb{N}}$ be a residual chain for Γ, and let $k \in \mathbb{N}$. Then*

$$b_k^{(2)}(X) = \lim_{n\to\infty} \frac{b_k(X_n)}{[\Gamma : \Gamma_n]}.$$

Here, X_n denotes the finite covering of X associated with the subgroup $\Gamma_n \subset \Gamma$ and b_k is the ordinary k-th $\mathbb{Q}$-Betti number (which equals the $\mathbb{C}$-Betti number).

Corollary 3.1.4 (approximation theorem, for groups). *Let Γ be a residually finite group of finite type, let $(\Gamma_n)_{n\in\mathbb{N}}$, and let $k \in \mathbb{N}$. Then*

$$b_k^{(2)}(\Gamma) = \lim_{n\to\infty} \frac{b_k(\Gamma_n)}{[\Gamma : \Gamma_n]}.$$

3.2 Proof of the approximation theorem

The proof of the approximation theorem (Theorem 3.1.3) is of a spectral nature; we will roughly follow unpublished notes of Sauer.

3.2.1 Reduction to kernels of self-adjoint operators

As first step, we reduce the approximation theorem to a statement about von Neumann dimensions of kernels of self-adjoint operators:

- On the one hand, if $\Lambda \curvearrowright Y$ is a free Λ-CW-complex (with a countable group Λ), then the (combinatorial) Laplacian Δ_* of the ℓ^2-chain complex of $\Lambda \curvearrowright Y$ is a positive self-adjoint operator on a Hilbert Λ-module that satisfies (Exercise 3.E.2)

$$b_k^{(2)}(\Lambda \curvearrowright Y) = \dim_{N\Lambda} \ker \Delta_k.$$

- On the other hand, the right-hand side in the approximation theorem (Theorem 3.1.3), can also be written as a von Neumann dimension: For each $n \in \mathbb{N}$, the finite group Γ/Γ_n is of finite type (Example 2.1.4), we have (Example 1.2.2)

$$\frac{b_k(X_n)}{[\Gamma : \Gamma_n]} = b_k^{(2)}(\Gamma/\Gamma_n \curvearrowright X_n),$$

and the boundary operator [Laplacian] on $C_*^{(2)}(\Gamma/\Gamma_n \curvearrowright X_n)$ is the reduction of the boundary operator [Laplacian] on $C_*^{(2)}(\Gamma \curvearrowright X)$ modulo Γ_n.

Therefore, Theorem 3.1.3 is a consequence of the following, slightly more algebraically looking, version (because the cellular Laplacian is defined over the integral group ring):

Theorem 3.2.1 (approximation theorem for kernels). *Let Γ be a finitely generated residually finite group with a residual chain $(\Gamma_n)_{n \in \mathbb{N}}$, let $m \in \mathbb{N}$, and let $A \in M_{m \times m}(\mathbb{Z}\Gamma)$ be self-adjoint and positive. Then*

$$\dim_{N\Gamma} \ker\big(M_A \colon (\ell^2\Gamma)^m \to (\ell^2\Gamma)^m\big)$$
$$= \lim_{n \to \infty} \dim_{N(\Gamma/\Gamma_n)} \ker\big(M_{A_n} \colon (\ell^2(\Gamma/\Gamma_n))^m \to (\ell^2(\Gamma/\Gamma_n))^m\big),$$

where $A_n \in M_{m \times m}(\mathbb{Z}[\Gamma/\Gamma_n])$ denotes the reduction of A modulo Γ_n.

We will now prove Theorem 3.2.1.

3.2.2 Reformulation via spectral measures

We reformulate the claim of Theorem 3.2.1 in terms of spectral measures: Let μ_A be the spectral measure on $\mathbb{R}$ (with the Borel σ-algebra) of the self-adjoint operator A. This measure has the following properties [23, Chapter 6][84, Chapter 5.2]:

- The measure μ_A is supported on the compact set $[0, a]$, where $a := \|A\|$.

- If $f : [0, \infty] \longrightarrow \mathbb{R}$ is a measurable bounded function, then the bounded linear operator $f(M_A)$, defined by functional calculus, satisfies

$$\int_{\mathbb{R}} f \, d\mu_A = \operatorname{tr}_\Gamma f(M_A).$$

- In the same way, for each $n \in \mathbb{N}$, the spectral measure μ_{A_n} of the reduction A_n of A is also supported on $[0, a]$ (because $\|A_n\| \leq \|A\|$).

Therefore, we obtain

$$\dim_{N\Gamma} \ker M_A = \operatorname{tr}_\Gamma(\text{orthogonal projection onto } \ker M_A)$$
$$= \operatorname{tr}_\Gamma \big(\chi_{\{0\}}(M_A)\big)$$
$$= \mu_A(\{0\})$$

and, for all $n \in \mathbb{N}$,

$$\dim_{N(\Gamma/\Gamma_n)} \ker M_{A_n} = \mu_{A_n}(\{0\}).$$

Hence, the claim of Theorem 3.2.1 is equivalent to the following property of the spectral measures: $\mu_A(\{0\}) = \lim_{n\to\infty} \mu_{A_n}(\{0\})$. We will now prove this statement on spectral measures.

3.2.3   Weak convergence of spectral measures

We first establish weak convergence of the spectral measures (Definition A.2.1):

Lemma 3.2.2 (weak convergence of spectral measures). *In this situation, the sequence $(\mu_{A_n})_{n\in\mathbb{N}}$ of measures on $\mathbb{R}$ weakly converges to μ_A, i.e., for all continuous functions $f : [0, a] \longrightarrow \mathbb{R}$, we have*

$$\lim_{n\to\infty} \int_{\mathbb{R}} f \, d\mu_{A_n} = \int_{\mathbb{R}} f \, d\mu_A.$$

Proof. The measures μ_{A_n} with $n \in \mathbb{N}$ and the measure μ_A are all supported on a common compact set (namely $[0, a]$). Therefore, by the Weierstraß approximation theorem, it suffices to take test functions of the form $(x \mapsto x^d)$ with $d \in \mathbb{N}$.

Thus, let $d \in \mathbb{N}$ and $f := (x \mapsto x^d)$. We then have

$$\int_{\mathbb{R}} f \, d\mu_A = \operatorname{tr}_\Gamma f(M_A) = \operatorname{tr}_\Gamma(A^d)$$

and $\int_{\mathbb{R}} f \, d\mu_{A_n} = \text{tr}_{\Gamma/\Gamma_n} A_n^d$.

Let $F \subset \Gamma$ be the support of A^d (i.e., the set of all elements of Γ that occur with non-zero coefficient in A). Because F is finite and $(\Gamma_n)_{n\in\mathbb{N}}$ is a residual chain, there exists an $N \in \mathbb{N}$ with

$$\forall_{n\in\mathbb{N}_{\geq N}} \quad F \cap \Gamma_n \subset \{e\}.$$

Then (by definition of the trace; check!)

$$\text{tr}_{\Gamma/\Gamma_n} A_n^d = \text{tr}_\Gamma A^d$$

for all $n \in \mathbb{N}_{\geq N}$. This shows weak convergence. $\square$

3.2.4   Convergence at 0

In general, weak convergence does *not* imply convergence of the measures on $\{0\}$ (Exercise 3.E.3). But by the portmanteau theorem (Theorem A.2.2), we at least obtain the following inequalities from Lemma 3.2.2:

$$\limsup_{n\to\infty} \mu_{A_n}(\{0\}) \leq \mu_A(\{0\})$$
$$\forall_{\varepsilon\in\mathbb{R}_{>0}} \quad \liminf_{n\to\infty} \mu_{A_n}((-\varepsilon,\varepsilon)) \geq \mu_A((-\varepsilon,\varepsilon)).$$

The first inequality already gives $\limsup_{n\to\infty} 1/[\Gamma_n : \Gamma] b_k(X_n) \leq b_k^{(2)}(X)$, the *Kazhdan inequality* [83].

In order to show the missing lower bound $\liminf_{n\to\infty} \mu_{A_n}(\{0\}) \geq \mu(\{0\})$, we will use integrality of the coefficients of A. More precisely, we will show:

Lemma 3.2.3. *In this situation, for all $n \in \mathbb{N}$ and all $\varepsilon \in (0,1)$, we have*

$$\mu_{A_n}((0,\varepsilon)) \leq \frac{m \cdot \ln(C)}{|\ln(\varepsilon)|},$$

where $C := \max(\|A\|, 1)$ (which does not depend on n).

Proof. Let $n \in \mathbb{N}$ and let $d := [\Gamma : \Gamma_n]$. Computing $\mu_{A_n}((0,\varepsilon))$ amounts to counting eigenvalues. We can view A_n as a matrix in $M_{m\times m}(\mathbb{Z}[\Gamma/\Gamma_n]) \subset M_{d\cdot m\times d\cdot m}(\mathbb{Z})$. In this view, A_n is symmetric and positive semi-definite (check!); let

$$0 = \lambda_1 \leq \cdots \leq \lambda_z = 0 < \lambda_{z+1} \leq \cdots \leq \lambda_{d\cdot m}$$

be the eigenvalues of A_n (listed with multiplicities). Then the characteristic polynomial of A_n is of the form $T^z \cdot q$ with $q \in \mathbb{Z}[T]$. In particular, $q(0) \neq 0$ and thus (because of integrality!)

$$\lambda_{z+1} \cdot \cdots \cdot \lambda_{d\cdot m} = |q(0)| \geq 1.$$

For $\varepsilon \in (0,1)$, let $M(\varepsilon)$ be the number of eigenvalues of A_n in $(0,\varepsilon)$. Then

$$1 \leq \lambda_{z+1} \cdot \cdots \cdot \lambda_{d \cdot m} \leq \varepsilon^{M(\varepsilon)} \cdot \|M_{A_n}\|^{d \cdot m - z - M(\varepsilon)} \leq \varepsilon^{M(\varepsilon)} \cdot C^{d \cdot m},$$

and so $M(\varepsilon) \leq d \cdot m \cdot \ln C / |\ln \varepsilon|$. Therefore, we obtain

$$\begin{aligned}
\mu_{A_n}\big((0,\varepsilon)\big) &= \dim_{N(\Gamma/\Gamma_n)} \text{ all eigenspaces of } A_n \text{ for eigenvalues in } (0,\varepsilon) \\
&= \frac{M(\varepsilon)}{d} \\
&\leq \frac{m \cdot \ln C}{|\ln(\varepsilon)|}.
\end{aligned}$$
$\square$

We can now complete the proof of Theorem 3.2.1 as follows: We have

$$\begin{aligned}
\liminf_{n \to \infty} \mu_{A_n}(\{0\}) &= \liminf_{n \to \infty} \big(\mu_{A_n}([0,\varepsilon]) - \mu_{A_n}((0,\varepsilon))\big) \\
&\geq \liminf_{n \to \infty} \mu_{A_n}\big((-\varepsilon,\varepsilon)\big) - \frac{m \cdot \ln C}{|\ln(\varepsilon)|} \qquad \text{(Lemma 3.2.3)} \\
&\geq \mu_A\big((-\varepsilon,\varepsilon)\big) - \frac{m \cdot \ln C}{|\ln(\varepsilon)|} \qquad \text{(portmanteau theorem)} \\
&\geq \mu_A(\{0\}) - \frac{m \cdot \ln C}{|\ln(\varepsilon)|}.
\end{aligned}$$

Taking $\varepsilon \to 0$ yields the desired estimate $\liminf_{n \to \infty} \mu_{A_n}(\{0\}) \geq \mu_A(\{0\})$. This finishes the proof of Theorem 3.2.1 and whence also of the approximation theorem (Theorem 3.1.3).

3.3 Homological gradient invariants

If I is a numerical invariant of (finitely generated residually finite) groups, then one can consider the associated *gradient invariant* $\widehat{I}$: If Γ is a finitely generated residually finite group and Γ_* is a residual chain in Γ, then

$$\widehat{I}(\Gamma, \Gamma_*) := \lim_{n \to \infty} \frac{I(\Gamma_n)}{[\Gamma : \Gamma_n]}.$$

This raises the following questions:

- Does the limit exist?

- Does $\widehat{I}(\Gamma, \Gamma_*)$ depend on the residual chain Γ_* of Γ?

- Does $\widehat{I}$ have a different interpretation?

3.3.1 Betti number gradients

For the gradient invariant associated to the ordinary Betti numbers, the approximation theorem (Theorem 3.1.3) gives a satisfying answer for finitely presented residually finite groups (of finite type).

Caveat 3.3.1. There exist finitely *generated* residually finite groups Γ with a residual chain $(\Gamma_n)_{n \in \mathbb{N}}$ such that the limit $\lim_{n \to \infty} b_1(\Gamma_n)/[\Gamma : \Gamma_n]$ does *not* exist [50].

In Chapter 5, we will discuss convergence of ordinary Betti numbers when moving from residual chains to BS-convergent sequences. Generalisations of the more classical version of the approximation theorem are surveyed in the literature [106, 91].

For $\mathbb{F}_p$-Betti number gradients the situation is much less understood. There are known positive examples of convergence/independence, but good candidates for alternative interpretations of the limits are rare.

3.3.2 Rank gradient

A non-commutative version of the first Betti number gradient is the rank gradient, introduced by Lackenby [93]:

Definition 3.3.2 (rank gradient). Let Γ be a finitely generated infinite residually finite group.

- For a finitely generated group Λ, we write $d(\Lambda)$ for the minimal size of a generating set of Λ.

- If Γ_* is a residual chain, then we define the *rank gradient of Γ with respect to Γ_** by
$$\mathrm{rg}(\Gamma, \Gamma_*) := \lim_{n \to \infty} \frac{d(\Gamma_n) - 1}{[\Gamma : \Gamma_n]}.$$

- Moreover, the *(absolute) rank gradient* of Γ is defined as
$$\mathrm{rg}\,\Gamma := \inf_{\Lambda \in \mathrm{F}(\Gamma)} \frac{d(\Lambda) - 1}{[\Gamma : \Lambda]},$$

where $\mathrm{F}(\Gamma)$ denotes the set of all finite index subgroups of Γ.

Remark 3.3.3. If Γ is a finitely generated group and $\Lambda \subset \Gamma$ is a finite index subgroup, then the rank estimate of the Nielsen–Schreier theorem shows that

$$d(\Lambda) - 1 \leq [\Gamma : \Lambda] \cdot \big(d(\Gamma) - 1\big).$$

Hence, the limit in the definition of the rank gradient indeed exists (and is equal to the infimum).

Remark 3.3.4 (rank gradient via normal subgroups). Let Γ be a finitely generated infinite group. Then

$$\operatorname{rg}\Gamma = \inf_{\Lambda\in F(\Gamma)} \frac{d(\Lambda) - 1}{[\Gamma : \Lambda]} = \inf_{\Lambda\in NF(\Gamma)} \frac{d(\Lambda) - 1}{[\Gamma : \Lambda]},$$

where $NF(\Gamma)$ denotes the set of all finite index normal subgroups of Γ (because every finite index subgroup of Γ contains a finite index subgroup that is normal in Γ).

Corollary 3.3.5 (rank gradient estimate for the first L^2-Betti number). *Let Γ be a finitely presented infinite residually finite group (of finite type). Then*

$$b_1^{(2)}(\Gamma) \leq \operatorname{rg}\Gamma.$$

Proof. If Λ is a finitely generated group, then there exists a classifying space of Λ with $d(\Lambda)$ one-dimensional cells. Therefore, $b_1(\Lambda) \leq d(\Lambda)$. Applying the approximation theorem (Theorem 3.1.3) to a residual chain Γ_* of Γ shows that

$$b_1^{(2)}(\Gamma) = \lim_{n\to\infty} \frac{b_1(\Gamma_n)}{[\Gamma : \Gamma_n]} \leq \lim_{n\to\infty} \frac{d(\Gamma_n)}{[\Gamma : \Gamma_n]} = \lim_{n\to\infty} \frac{d(\Gamma_n) - 1}{[\Gamma : \Gamma_n]} = \operatorname{rg}(\Gamma, \Gamma_*).$$

Taking the infimum over all residual chains, we obtain with Remark 3.3.3 and Remark 3.3.4 that

$$b_1^{(2)}(\Gamma) \leq \inf_{\Lambda\in NF(\Gamma)} \frac{d(\Lambda) - 1}{[\Gamma : \Lambda]} = \operatorname{rg}\Gamma.$$

Alternatively, one can also show directly that $b_1^{(2)}(\Lambda) \leq d(\Lambda)$ for all finitely generated groups and then use multiplicativity of $b_1^{(2)}$ under finite coverings (Theorem 2.2.2). $\qquad\square$

It remains an open problem to determine whether the rank gradient depends on the residual chain; in all known cases, the inequality in Corollary 3.3.5 is an equality and the absolute rank gradient can be computed by every residual chain. This is related to the fixed price problem (Outlook 4.3.13).

The Betti number-rank estimate can be improved to estimates for the minimal size of normal generating sets in terms of the first L^2-Betti number; this gives lower bounds on the girth of certain Cayley graphs [131, Theorem 5.1].

3.3.3 More gradients

Further examples of gradient invariants are:

- Homology log-torsion gradients (which conjecturally might be related to L^2-torsion?!) [106].

- Simplicial volume gradients (Chapter 6).

3.E Exercises

Exercise 3.E.1 (surface groups, free groups). Prove the approximation theorem for surface groups and free groups by direct computation of the right-hand side. Compute the (absolute) rank gradient of surface groups and free groups.

Exercise 3.E.2 (Laplacian [105, Lemma 1.18]). Let Γ be a countable group, let C_* be a chain complex of Hilbert Γ-modules (with boundary operators ∂_*), and let Δ_* be the *Laplacian* of C_*, i.e., for each $n \in \mathbb{N}$, we set

$$\Delta_n := \partial_{n+1} \circ \partial_{n+1}^* + \partial_n^* \circ \partial_n.$$

Show that there exists an isomorphism

$$\ker \Delta_n \longrightarrow \ker \partial_n \,/\, \overline{\operatorname{im} \partial_{n+1}}$$

of Hilbert Γ-modules.
Hints. Consider the orthogonal projection onto $\overline{\operatorname{im} \partial_{n+1}}^{\perp}$.

Exercise 3.E.3 (weak convergence). Give an example of a sequence $(\mu_n)_{n \in \mathbb{N}}$ of probability measures on $[0,1]$ (with the Borel σ-algebra) that weakly converges to a probability measure μ on $[0,1]$, but that satisfies

$$\lim_{n \to \infty} \mu_n(\{0\}) \neq \mu(\{0\}).$$

Exercise 3.E.4 (rank gradients of products). Let Γ and Λ be finitely generated infinite residually finite groups. Compute $\operatorname{rg}(\Gamma \times \Lambda)$.

Exercise 3.E.5 (self-maps). Let M be an oriented closed connected aspherical manifold with residually finite fundamental group Γ. Moreover, we suppose that M admits a self-map $f \colon M \longrightarrow M$ with $|\deg f| \geq 2$.

1. Give examples of this situation.

2. Show that $\operatorname{rg}(\Gamma) = 0$.

3. Show that $b_k^{(2)}(\Gamma) = 0$ for all $k \in \mathbb{N}$ [105, Theorem 14.40].

4. Challenge: Does the vanishing of L^2-Betti numbers of Γ also hold without any residual finiteness or Hopficity condition on Γ ?! (This is an open problem.)

Hints. Covering theory shows that $\operatorname{im} f$ has finite index in Γ. Moreover, it is useful to know that residually finite groups are *Hopfian*, i.e., every self-epimorphism is an automorphism.

4

The dynamical view: Measured group theory

The theory of von Neumann algebras can be viewed as a model of non-commutative measure theory. Therefore, it is plausible that L^2-Betti numbers can be computed in terms of probability measure preserving actions.

- On the one hand, this leads to an additional way of computing L^2-Betti numbers of groups.

- On the other hand, in this way, L^2-Betti numbers provide orbit equivalence invariants.

We will first recall some basic terminology from measured group theory. Then we will study L^2-Betti numbers of standard equivalence relations. Moreover, we will discuss cost and its relation with the first L^2-Betti number and rank gradients.

Overview of this chapter.

4.1	Measured group theory	38
4.2	L^2-Betti numbers of equivalence relations	40
4.3	Cost of groups	47
4.E	Exercises	57

Running example. amenable groups, free groups

4.1 Measured group theory

Measured group theory is the theory of dynamical systems, i.e., of (probability) measure preserving actions of groups. We briefly introduce some of the terminology. More information can be found in the literature [62, 65, 89, 90].

4.1.1 Standard actions

Definition 4.1.1 (standard action). Let Γ be a countable group.

- A *standard action of* Γ is an action of Γ on a standard Borel probability space by measure preserving Borel automorphisms.

- A standard action $\Gamma \curvearrowright (X, \mu)$ is *essentially free* if μ-almost every point has trivial stabiliser group.

- A standard action $\Gamma \curvearrowright (X, \mu)$ is *ergodic* if every measurable subset $A \subset X$ with $\Gamma \cdot A = A$ satisfies $\mu(A) \in \{0, 1\}$.

A *standard Borel space* is a measurable space that is isomorphic to a Polish space with its Borel σ-algebra. Standard Borel spaces form a convenient category for measure theory [87].

Example 4.1.2 (Bernoulli shift). Let Γ be a countable group. Then the shift action of Γ on the product space $\prod_\Gamma \{0, 1\}$ (with the product σ-algebra and the product of the uniform distribution on $\{0, 1\}$) is a standard action of Γ.

Moreover, if Γ is infinite, this action is essentially free and ergodic [126, Lemma 3.37][90, Chapter 2.3.1].

Example 4.1.3 (finite quotients). Let Γ be a countable group and let $\Lambda \subset \Gamma$ be a finite index subgroup. Then the translation action of Γ on the coset space Γ/Λ (with the discrete σ-algebra and the uniform distribution) is a standard action. It is ergodic, but apart from pathological cases, not essentially free.

Example 4.1.4 (profinite completion). Let Γ be a finitely generated group. We then consider the *profinite completion of* Γ, defined by

$$\widehat{\Gamma} := \varprojlim_{\Lambda \in \mathrm{NF}(\Gamma)} \Gamma/\Lambda,$$

where $\mathrm{NF}(\Gamma)$ denotes the set of all finite index normal subgroups of Γ. Then $\widehat{\Gamma}$ is a group with the induced composition and the diagonal map $\Gamma \longrightarrow \widehat{\Gamma}$ is a group homomorphism, which leads to a Γ-action on $\widehat{\Gamma}$ (by translation of each component). The group Γ is residually finite if and only if this action on $\widehat{\Gamma}$ is free (Exercise 4.E.1).

Moreover, we can equip $\widehat{\Gamma}$ with the inverse limit of the discrete σ-algebras and the inverse limit of the uniform probability measures on the finite factors. This action $\Gamma \curvearrowright \widehat{\Gamma}$ is then a standard action.

4.1.2 Measure/orbit equivalence

We will now compare measure preserving actions of groups via couplings and their orbit structure, respectively. Measure equivalence is a measure-theoretic version of quasi-isometry (the connection being given by Gromov's topological criterion for quasi-isometry [74, 0.2.C_2'])

Definition 4.1.5 (measure equivalence [74, 0.5.E]). Let Γ and Λ be countable infinite groups.

- An *ME coupling* between the groups Γ and Λ is a standard Borel measure space (Ω, μ) of infinite measure together with a measure preserving action of $\Gamma \times \Lambda$ by Borel automorphisms so that both actions $\Gamma \curvearrowright (\Omega, \mu)$ and $\Lambda \curvearrowright (\Omega, \mu)$ admit fundamental domains Y and X, respectively, of finite measure. The *index* of such an ME coupling is the quotient $\mu(X)/\mu(Y)$.

- The groups Γ and Λ are *measure equivalent* if there exists an ME coupling between them; in this case, we write $\Gamma \sim_{\mathrm{ME}} \Lambda$.

Example 4.1.6 (lattices are measure equivalent). Let G be a locally compact second countable group (with infinite Haar measure μ) and let $\Gamma, \Lambda \subset G$ be lattices in G. Then the action

$$(\Gamma \times \Lambda) \times G \longrightarrow G$$
$$((\gamma, \lambda), x) \longmapsto \gamma \cdot x \cdot \lambda^{-1}$$

shows that the ambient group G yields an ME coupling between Γ and Λ.

In particular, countable infinite commensurable groups are measure equivalent. For instance, $F_n \sim_{\mathrm{ME}} F_m$ for all $n, m \in \mathbb{N}_{\geq 2}$.

Measure equivalence indeed defines an equivalence relation on the class of all countable infinite groups [62, p. 300].

Definition 4.1.7 ((stable) orbit equivalence). Let $\Gamma \curvearrowright X$ and $\Lambda \curvearrowright Y$ be standard actions.

- The actions $\Gamma \curvearrowright X$ and $\Lambda \curvearrowright Y$ are *orbit equivalent* if there exists a measure preserving Borel isomorphism $f\colon X' \longrightarrow Y'$ between co-null subsets $X' \subset X$ and $Y' \subset Y$ with

$$\forall_{x \in X'} \quad f(\Gamma \cdot x \cap X') = \Lambda \cdot f(x) \cap Y'.$$

In this case, we write $\Gamma \curvearrowright X \sim_{OE} \Lambda \curvearrowright Y$.

- The actions $\Gamma \curvearrowright X$ and $\Lambda \curvearrowright Y$ are *stably orbit equivalent* if there exists a Borel isomorphism $f \colon X' \longrightarrow Y'$ between measurable subsets $X' \subset X$ and $Y' \subset Y$ with $\mu(X') > 0$, $\mu(Y') > 0$ that satisfies $1/\mu(X') \cdot f_* \mu|_{X'} = 1/\nu(Y') \cdot \nu|_{Y'}$ with

$$\forall_{x \in X'} \quad f(\Gamma \cdot x \cap X') = \Lambda \cdot f(x) \cap Y'.$$

The *index* of such a stable orbit equivalence f is $\mu(Y')/\mu(X')$ and we write $\Gamma \curvearrowright X \sim_{SOE} \Lambda \curvearrowright Y$.

Moreover, we call Γ and Λ [stably] orbit equivalent if they admit [stably] orbit equivalent standard actions.

Theorem 4.1.8 (measure equivalence [62, Theorem 2.5]). *Two countable groups are measure equivalent if and only if they admit essentially free standard actions that are stably orbit equivalent.*

Example 4.1.9 (lattices are stably orbit equivalent). In view of Theorem 4.1.8 and Example 4.1.6, we obtain: Lattices in locally compact second countable topological groups with infinite Haar measure are stably orbit equivalent.

In particular, $F_n \sim_{SOE} F_m$ for all $n, m \in \mathbb{N}_{\geq 2}$.

In order to get a better understanding of measure/orbit equivalence of groups, we need suitable invariants. A first example is amenability [118, 122]; the class of amenable groups contains, for instance, the class of all virtually solvable groups. Orbit/measure equivalence invariance of amenability is established by the theorems of Dye, Connes–Feldman-Weiss, and Ornstein–Weiss:

Theorem 4.1.10 (dynamical characterisation of amenable groups [114][89, Chapter 10][90, Chapter 4.8/4.9]).

1. *A countable infinite group is amenable if and only if it is measure equivalent to* $\mathbb{Z}$.

2. *Any two ergodic standard actions of any two countable infinite amenable groups are orbit equivalent.*

Further examples of suitable invariants are (vanishing of) L^2-Betti numbers, the (sign of the) Euler characteristic, and cost, as we will outline now.

4.2 L^2-Betti numbers of equivalence relations

Orbit equivalence is a notion that does not directly involve a group (action), but only the orbit equivalence relations of standard actions. It is therefore natural to widen the context by studying equivalence relations in this measured setting. Gaboriau discovered that one can define L^2-Betti numbers of such equivalence relations and how these relate to L^2-Betti numbers of groups [64].

4.2.1   Measured equivalence relations

Definition 4.2.1 (standard equivalence relation).

- A *standard equivalence relation* is an equivalence relation $\mathcal{R} \subset X \times X$ on a standard Borel space X with the following properties:

  - The subset $\mathcal{R} \subset X \times X$ is measurable.
  - Each $\mathcal{R}$-equivalence class is countable.

- If $\mathcal{R}$ is a standard equivalence relation on X and $A \subset X$ is a measurable subset, then we define the *restriction to A* by

$$\mathcal{R}|_A := \big\{ (x,y) \mid x,y \in A, \ (x,y) \in \mathcal{R} \big\} \subset A \times A.$$

- A *measured standard equivalence relation* is a standard equivalence relation $\mathcal{R}$ on a standard Borel probability space (X, μ) with the following property: Every partial $\mathcal{R}$-automorphism of X is μ-preserving. A *partial $\mathcal{R}$-automorphism* is a Borel automorphism $A \longrightarrow B$ between measurable subsets $A, B \subset X$ whose graph is contained in $\mathcal{R}$. The groupoid of partial automorphisms of $\mathcal{R}$ is denoted by $[\![\mathcal{R}]\!]$.

Example 4.2.2 (orbit relations). Let $\Gamma \curvearrowright X$ be a standard action. Then the *orbit relation*

$$\mathcal{R}_{\Gamma \curvearrowright X} := \big\{ (x, \gamma \cdot x) \mid x \in X, \ \gamma \in \Gamma \big\} \subset X \times X$$

is a measured standard equivalence relation. Conversely, every standard equivalence relation arises in this way, as shown by the Feldman–Moore theorem [56].

4.2.2 L^2-Betti numbers of equivalence relations

We now introduce L^2-Betti numbers of measured standard equivalence relations, following Sauer's "algebraic" approach [123] (and we will also add the references for proofs in this language). The original construction is due to Gaboriau and has a more simplicial/geometric flavour [64]; both constructions are known to result in the same numbers [112].

For the definition of L^2-Betti numbers of a group Γ (of finite type), we used the following ingredients:

- Base ring: The field $\mathbb{C}$.

- Extension by the group: The group ring $\mathbb{C}\Gamma$.

- Completion of scalars: The group von Neumann algebra $N\Gamma$.

- The trace/dimension on $N\Gamma$.

- Modules with a projectivity condition: Hilbert Γ-modules.

- A suitable model of the classifying space of Γ.

If $\mathcal{R}$ is a measured standard equivalence relation on (X, μ), we will use the corresponding replacements listed below; for the definition of the dimension function, we will work in the extended algebraic setting, which we sketched in Outlook 1.2.4 for the group case.

- Base ring: The function space $L^\infty(X) := L^\infty(X, \mathbb{C})$.

- Extension by the equivalence relation: The ring

$$\mathbb{C}\mathcal{R} := \big\{ f \in L^\infty(\mathcal{R}, \nu) \ \big| \ \sup_{x \in X} \big|\{y \mid f(x, y) \neq 0\}\big| < \infty,$$

$$\sup_{y \in X} \big|\{x \mid f(x, y) \neq 0\}\big| < \infty \big\}$$

with the "convolution" product $(f \cdot g)(x, y) := \sum_{z \in [x]_{\mathcal{R}}} f(x, z) \cdot g(z, y)$. Here, we use the following measure on $\mathcal{R}$:

$$\nu \colon \text{Borel } \sigma\text{-algebra on } \mathcal{R} \longrightarrow \mathbb{R}_{\geq 0}$$

$$A \longmapsto \int_X \big|A \cap (\{x\} \times X)\big| \, d\mu(x, y).$$

- Completion of scalars: The von Neumann algebra $N\mathcal{R}$ is the weak closure of $\mathbb{C}\mathcal{R}$ in $B(L^2(\mathcal{R}, \nu))$ with respect to the right convolution action of $\mathbb{C}\mathcal{R}$.

- The von Neumann algebra $N\mathcal{R}$ also admits a trace:

$$\text{tr}_{\mathcal{R}} \colon N\mathcal{R} \longrightarrow \mathbb{C}$$

$$a \longmapsto \langle \chi_\Delta, a(\chi_\Delta) \rangle$$

- Dimension function: If P is a finitely generated projective $N\mathcal{R}$-module, we set

$$\text{pdim}_{N\mathcal{R}} P := \text{tr}_{\mathcal{R}} p = \text{tr}_{\mathcal{R}} A = \sum_{j=1}^{n} \text{tr}_{\mathcal{R}} A_{jj} \in \mathbb{R}_{\geq 0},$$

where $p\colon (N\mathcal{R})^n \longrightarrow (N\mathcal{R})^n$ is a projection with $P \cong_{N\mathcal{R}} \operatorname{im} p$ and associated matrix $A \in M_{n\times n}(N\mathcal{R})$. More generally, for an $N\mathcal{R}$-module V, one sets

$$\dim_{N\mathcal{R}} V := \sup\{\operatorname{pdim}_{N\mathcal{R}} P \mid P \text{ is a finitely generated projective}$$
$$N\mathcal{R}\text{-submodule of } V\} \in \mathbb{R}_{\geq 0} \cup \{\infty\}.$$

As in Outlook 1.2.4, this leads to a reasonable notion of dimension because the module theory of $N\mathcal{R}$ is well-behaved: the ring $N\mathcal{R}$ is semi-hereditary.

- Instead of a classifying space (as in Gaboriau's work), we use the algebraic description of group homology as a Tor-functor.

Definition 4.2.3 (L^2-Betti numbers of equivalence relations). Let $\mathcal{R}$ be a measured standard equivalence relation on (X, μ) and let $n \in \mathbb{N}$. Then the *n-th L^2-Betti number of $\mathcal{R}$* is defined by

$$b_n^{(2)}(\mathcal{R}) := \dim_{N\mathcal{R}} \operatorname{Tor}_n^{\mathbb{C}\mathcal{R}}(N\mathcal{R}, L^\infty(X)) \in \mathbb{R}_{\geq 0} \cup \{\infty\}.$$

The spatial restriction of measured equivalence relations leads to scaled L^2-Betti numbers:

Theorem 4.2.4 (restriction formula [64, Corollaire 5.5][123]). *Let $\mathcal{R}$ be a measured standard equivalence relation on (X, μ) and let $A \subset X$ be a measurable subset with $\mu(A) > 0$. Then, for all $n \in \mathbb{N}$,*

$$b_n^{(2)}(\mathcal{R}|_A) = \frac{1}{\mu(A)} \cdot b_n^{(2)}(\mathcal{R}).$$

4.2.3 Comparison with L^2-Betti numbers of groups

The key observation is that L^2-Betti numbers of orbit relations coincide with the L^2-Betti numbers of the group:

Theorem 4.2.5 (L^2-Betti numbers of groups vs. equivalence relations [64, Corollaire 3.16]). *Let Γ be a group of finite type and let $\Gamma \curvearrowright X$ be an essentially free standard action of Γ. Then, for all $n \in \mathbb{N}$,*

$$b_n^{(2)}(\Gamma) = b_n^{(2)}(\mathcal{R}_{\Gamma \curvearrowright X}).$$

For the proof, we will follow Sauer's approach [123]. As a first step, we introduce an intermediate object:

Remark 4.2.6 (the crossed product ring). Let Γ be a countable group and let $\Gamma \curvearrowright X$ be an essentially free standard action. In addition to the group ring $\mathbb{C}\Gamma$ and the equivalence relation ring $\mathbb{C}\mathcal{R}_{\Gamma \curvearrowright X}$, we also have the *algebraic* *crossed product* $L^\infty(X) \rtimes \Gamma$, where $g \in \Gamma$ acts by $f \mapsto (x \mapsto f(g^{-1} \cdot x))$ on $L^\infty(X)$. More explicitly, the underlying $\mathbb{C}$-vector space is $\bigoplus_\Gamma L^\infty(X)$; we denote the element in the summand indexed by $\gamma \in \Gamma$ corresponding to $f \in L^\infty(X)$ by $f[\gamma]$. Then the multiplication on $L^\infty(X) \rtimes \Gamma$ is given by

$$f[\gamma] \cdot f'[\gamma'] := (x \mapsto f(x) \cdot f'(\gamma^{-1} \cdot x))[\gamma \cdot \gamma']$$

for all $f, f' \in L^\infty(X)$ and all $\gamma, \gamma' \in \Gamma$.

On the one hand, the inclusion of $\mathbb{C}$ into $L^\infty(X)$ via constant functions leads to a ring inclusion $\mathbb{C}\Gamma \hookrightarrow L^\infty(X) \rtimes \Gamma$. The ring $L^\infty(X) \rtimes \Gamma$ is flat as a right $\mathbb{C}\Gamma$-module, because for all $\mathbb{C}\Gamma$-modules V, we have a canonical isomorphism (check!)

$$(L^\infty(X) \rtimes \Gamma) \otimes_{\mathbb{C}\Gamma} V \cong_\mathbb{C} L^\infty(X) \otimes_\mathbb{C} V$$

and the tensor product $L^\infty(X) \otimes_\mathbb{C} \cdot$ is exact.

On the other hand, we can view $L^\infty(X) \rtimes \Gamma$ as a subring of $\mathbb{C}\mathcal{R}_{\Gamma \curvearrowright X}$ (check!):

$$L^\infty(X) \rtimes \Gamma \longrightarrow \mathbb{C}\mathcal{R}_{\Gamma \curvearrowright X}$$
$$f[\gamma] \longmapsto \left((x,y) \mapsto \begin{cases} f(x) & \text{if } y = \gamma^{-1} \cdot x \\ 0 & \text{otherwise} \end{cases} \right)$$

Moreover, we will need some algebraic facts on finite von Neumann algebras and trace-preserving $*$-homomorphisms:

Remark 4.2.7 (trace-preserving $*$-homomorphisms). A *finite von Neumann algebra* is a von Neumann algebra that admits a faithful finite normal trace [57, Chapter 4.8]. Examples are the group von Neumann algebras of countable groups, the group von Neumann algebras of measured standard equivalence relations or the spaces $L^\infty(X)$ of standard Borel probability spaces X (with the trace given by integration).

If $f \colon A \longrightarrow B$ is a trace-preserving $*$-homomorphism between finite von Neumann algebras, then we have [123]:

- The ring homomorphism f is faithfully flat and

- dimension-preserving: For all left A-modules V, we obtain

$$\dim_B(B \otimes_A V) = \dim_A V,$$

where the right A-module structure on B is defined via f.

An example of such a trace-preserving $*$-homomorphism between finite von Neumann algebras is the canonical inclusion $N\Gamma \longrightarrow N\mathcal{R}_{\Gamma \curvearrowright X}$ induced by an essentially free standard action $\Gamma \curvearrowright X$ (check!).

Proof of Theorem 4.2.5. We abbreviate $\mathcal{R} := \mathcal{R}_{\Gamma \curvearrowright X}$. In view of Remark 4.2.6 and Remark 4.2.7, we have the following commutative diagram of rings (where all unmarked arrows denote canonical inclusions).

$$
\begin{array}{ccc}
\mathbb{C} \longrightarrow L^\infty(X) = L^\infty(X) \\
\downarrow \qquad\qquad \downarrow \qquad\qquad \downarrow \\
\mathbb{C}\Gamma \longrightarrow L^\infty(X) \rtimes \Gamma \longrightarrow \mathbb{C}\mathcal{R} \\
\downarrow \qquad\qquad\qquad\qquad\qquad \downarrow \\
N\Gamma \dashrightarrow N\mathcal{R}
\end{array}
$$
$$\text{trace-preserving *-homomorphism}$$

We can then perform the following computation [123]:

$b_n^{(2)}(\Gamma) = \dim_{N\Gamma}$ (reduced n-th $\ell^2\Gamma$-homology of a finite type model of $B\Gamma$)
  (by definition)

$= \dim_{N\Gamma}$ (algebraic n-th $N\Gamma$-homology of a finite type model of $B\Gamma$)
  (by the properties of the extended dimension [105, Lemma 6.53])

$= \dim_{N\Gamma} \operatorname{Tor}_n^{\mathbb{C}\Gamma}(N\Gamma, \mathbb{C})$
  (by the geometric computation of Tor)

$= \dim_{N\mathcal{R}} \big(N\mathcal{R} \otimes_{N\Gamma} \operatorname{Tor}_n^{\mathbb{C}\Gamma}(N\Gamma, \mathbb{C})\big)$
  (trace-pres. *-homs are dimension-preserving; Remark 4.2.7)

$= \dim_{N\mathcal{R}} \operatorname{Tor}_n^{\mathbb{C}\Gamma}(N\mathcal{R}, \mathbb{C})$
  (trace-pres. *-homs are faithfully flat; Remark 4.2.7)

$= \dim_{N\mathcal{R}} \operatorname{Tor}_n^{L^\infty(X)\rtimes\Gamma}\big(N\mathcal{R}, (L^\infty(X) \rtimes \Gamma) \otimes_{\mathbb{C}\Gamma} \mathbb{C}\big)$
  ($L^\infty(X) \rtimes \Gamma$ is flat as right $\mathbb{C}\Gamma$-module; Remark 4.2.6)

$= \dim_{N\mathcal{R}} \operatorname{Tor}_n^{L^\infty(X)\rtimes\Gamma}\big(N\mathcal{R}, L^\infty(X)\big)$
  (Remark 4.2.6)

We now need to replace $\operatorname{Tor}_n^{L^\infty(X)\rtimes\Gamma}$ with $\operatorname{Tor}_n^{\mathbb{C}\mathcal{R}}$. Indeed, both Tor-terms have the same $N\mathcal{R}$-dimension: The inclusion $L^\infty(X) \rtimes \Gamma \hookrightarrow \mathbb{C}\mathcal{R}$ is a $\dim_{L^\infty(X)}$-isomorphism and $\mathbb{C}\mathcal{R}$ is a so-called $L^\infty(X)$-$L^\infty(X)$-dimension compatible bimodule [123]. Carefully stepping through the homological algebra shows $\dim_{N\mathcal{R}} \operatorname{Tor}_n^{L^\infty(X)\rtimes\Gamma}(N\mathcal{R}, L^\infty(X)) = \dim_{N\mathcal{R}} \operatorname{Tor}_n^{\mathbb{C}\mathcal{R}}(N\mathcal{R}, L^\infty(X))$ [123]. Therefore, we obtain

$$b_n^{(2)}(\Gamma) = \dim_{N\mathcal{R}} \operatorname{Tor}_n^{\mathbb{C}\mathcal{R}}(N\mathcal{R}, L^\infty(X))$$
$$= b_n^{(2)}(\mathcal{R}),$$

as claimed. $\square$

4.2.4   Applications to orbit equivalence

Corollary 4.2.8 (OE/ME-invariants [64, Corollaire 5.6]). *Let Γ and Λ be infinite groups of finite type and let $n \in \mathbb{N}$.*

1. *If $\Gamma \curvearrowright X$ and $\Lambda \curvearrowright Y$ are orbit equivalent (essentially) free standard actions, then $b_n^{(2)}(\Gamma) = b_n^{(2)}(\Lambda)$.*

2. *If Γ and Λ are measure equivalent (with index c), then*

$$b_n^{(2)}(\Gamma) = c \cdot b_n^{(2)}(\Lambda).$$

In particular, $b_n^{(2)}(\Gamma)$ and $b_n^{(2)}(\Lambda)$ have the same vanishing behaviour.

Proof. This follows from Theorem 4.2.5 and 4.2.4 (Exercise 4.E.2).   □

Corollary 4.2.9 (non-orbit equivalence of free groups). *Let $n, m \in \mathbb{N}$ and let F_n and F_m be free groups of rank n and m, respectively. Then F_n and F_m admit orbit equivalent standard actions if and only if $n = m$.*

Proof. If F_n and F_m admit orbit equivalent standard actions, then $b_1^{(2)}(F_n) = b_1^{(2)}(F_m)$ (Corollary 4.2.8). Because of $b_1^{(2)}(F_n) = n - 1$ and $b_1^{(2)}(F_m) = m - 1$ (Example 2.2.6), we obtain $n = m$.   □

4.2.5   Applications to L^2-Betti numbers of groups

Corollary 4.2.10 (L^2-Betti numbers of amenable groups). *Let Γ be an amenable group of finite type and let $n \in \mathbb{N}$. Then*

$$b_n^{(2)}(\Gamma) = \begin{cases} \frac{1}{|G|} & \text{if } n = 0 \\ 0 & \text{if } n > 0. \end{cases}$$

In particular: If Γ admits a finite classifying space, then $\chi(\Gamma) = 0$.

Proof. This can be deduced from our previous computations, Theorem 4.1.10, and Corollary 4.2.8 (Exercise 4.E.4). The original proof of Cheeger and Gromov is based on a Følner-type argument [36].   □

Corollary 4.2.11 (proportionality principle for L^2-Betti numbers [64, Corollaire 0.2, Théorème 6.3]). *Let Γ, Λ be lattices in a locally compact second countable topological group G (with a given Haar measure) and let $n \in \mathbb{N}$. Then*

$$\frac{b_n^{(2)}(\Gamma)}{\mathrm{vol}(\Gamma \setminus G)} = \frac{b_n^{(2)}(\Lambda)}{\mathrm{vol}(\Lambda \setminus G)}.$$

This common quotient is denoted by $b_n^{(2)}(G)$.

Proof. Lattices in the same group are measure equivalent (Example 4.1.6) and the index is the ratio of covolumes. We then apply Corollary 4.2.8.

In the locally symmetric case, this kind of proportionality can also be obtained analytically via the heat kernel. □

It should be noted that L^2-Betti numbers for locally compact second countable groups can also be defined directly [119] and that the values depend on the choice of a Haar measure. Usually, in applications, such L^2-Betti numbers appear together with the covolume of a lattice (and so the dependence on the Haar measure is irrelevant).

4.3 Cost of groups

Cost of measured equivalence relations is a measure theoretic version of "minimal size of a generating set". More precisely:

Definition 4.3.1 (graphing, cost [96]). Let $\mathcal{R}$ be a measured equivalence relation on (X, μ).

- A *graphing* of $\mathcal{R}$ is a family $\Phi = (\varphi_i)_{i \in I}$ of partial $\mathcal{R}$-automorphisms of (X, μ) such that
$$\langle \Phi \rangle = \mathcal{R},$$
where $\langle \Phi \rangle$ denotes the minimal (with respect to inclusion) equivalence relation on X that contains the graphs of all φ_i with $i \in I$.

- The *cost* of a graphing Φ of $\mathcal{R}$ is
$$\text{cost}\,\Phi := \sum_{\varphi \in \Phi} \mu(\text{domain of } \varphi) \in \mathbb{R}_{\geq 0} \cup \{\infty\}.$$

- The *cost of* $\mathcal{R}$ is
$$\text{cost}\,\mathcal{R} := \inf\{\text{cost}\,\Phi \mid \Phi \text{ is a graphing of } \mathcal{R}\} \in \mathbb{R}_{\geq 0} \cup \{\infty\}.$$

More geometrically, one can also introduce graphings in the language of Borel graphs.

Definition 4.3.2 (cost of a group [63]). Let Γ be a countable group.

- If $\Gamma \curvearrowright X$ is a standard action, we write $\text{cost}(\Gamma \curvearrowright X) := \text{cost}(\mathcal{R}_{\Gamma \curvearrowright X})$.

- The *cost of* Γ is defined as
$$\text{cost}\,\Gamma := \inf\{\text{cost}(\Gamma \curvearrowright X) \mid \Gamma \curvearrowright X \text{ is a standard action}\} \in \mathbb{R}_{\geq 0} \cup \{\infty\}.$$

Remark 4.3.3 (on the definition of cost). The term on the right-hand side in the above definition can indeed be formalised as a proper set (every standard Borel space can be modelled on $[0,1]$ and thus we can form a *set* of representatives of all isomorphism classes of all standard Γ-actions). Moreover, taking products with an essentially free standard Γ-action shows that it suffices to consider essentially free actions and that the infimum is attained [89, Proposition 29.1]. Ergodic decomposition shows that, alternatively, it suffices to consider ergodic standard actions [89, Remark 29.2].

Remark 4.3.4 (the (trivial) rank estimate). If Γ is a group, then $\mathrm{cost}(\Gamma) \leq d(\Gamma)$: In each orbit equivalence relation, group elements yield (total) automorphisms and the automorphisms associated with a generating set clearly also generate the orbit equivalence relation.

4.3.1   Rank gradients via cost

The residually finite view and the dynamical view on the minimal number of generators are unified through the profinite completion (Example 4.1.4):

Theorem 4.3.5 (cost of the profinite completion [7]). *Let Γ be a finitely generated residually finite infinite group. Then*

$$\mathrm{rg}\,\Gamma = \mathrm{cost}(\Gamma \curvearrowright \widehat{\Gamma}) - 1.$$

Furthermore, if Γ_* is a residual chain in Γ, then (where $\widehat{\Gamma}_* := \varprojlim_{n\in\mathbb{N}} \Gamma/\Gamma_n$)

$$\mathrm{rg}(\Gamma, \Gamma_*) = \mathrm{cost}(\Gamma \curvearrowright \widehat{\Gamma}_*) - 1.$$

The original statement by Abért and Nikolov is slightly more general (the arguments work for Farber chains as well) and the original proof is formulated in terms of product cost.

Setup 4.3.6. Let Γ be a finitely generated residually finite infinite group. We write F for the family of subgroups of Γ in question (the family of all normal finite index subgroups of Γ or the given chain Γ_*), we write $X := \varprojlim_{\Lambda\in F} \Gamma/\Lambda$ for the corresponding profinite completion with probability measure μ, and for $\Lambda \in F$, we write $\pi_\Lambda \colon X \longrightarrow \Gamma/\Lambda$ for the associated structure map.

Proof. We assume the notation from Setup 4.3.6 and begin the proof with the (simple) estimate "$\geq$": Let $\Lambda \in F$, let $d := d(\Lambda)$, and $n := [\Gamma : \Lambda]$. Then, we choose a generating set $\{\lambda_1, \ldots, \lambda_d\}$ of Λ and coset representatives $\gamma_1, \ldots, \gamma_n$ for Γ/Λ. Moreover, we consider the "cylinder"

$$A := \pi_\Lambda^{-1}(e \cdot \Lambda) \subset X,$$

which has measure $\mu(A) = 1/[\Gamma : \Lambda]$, as well as the translation maps

$$\varphi_j := \lambda_j \cdot\; : A \longrightarrow \lambda_j \cdot A = A$$
$$\psi_k := \gamma_k \cdot\; : A \longrightarrow \gamma_k \cdot A$$

for $j \in \{1, \ldots, d\}$ and $k \in \{1, \ldots, n\}$. Then

$$\Phi := (\varphi_j)_{j \in \{1,\ldots,d\}} \cup (\psi_k)_{k \in \{1,\ldots,n\}}$$

is a subfamily of $[\![\mathcal{R}_{\Gamma \curvearrowright X}]\!]$ and a straightforward computation shows that Φ is a graphing of $\mathcal{R}_{\Gamma \curvearrowright X}$ (check!). Therefore, we obtain

$$\mathrm{cost}(\Gamma \curvearrowright X) \leq \mathrm{cost}(\Phi) = d \cdot \mu(A) + n \cdot \mu(A)$$

$$= d(\Lambda) \cdot \frac{1}{[\Gamma : \Lambda]} + [\Gamma : \Lambda] \cdot \frac{1}{[\Gamma : \Lambda]} = \frac{d(\Lambda)}{[\Gamma : \Lambda]} + 1.$$

Taking the infimum over all Λ in F shows that $\mathrm{cost}(\Gamma \curvearrowright X) - 1 \leq \mathrm{rg}(\Gamma, F)$.

We will now establish the converse estimate "$\leq$": Let Φ be a graphing of $\mathcal{R}_{\Gamma \curvearrowright X}$; as $\mathrm{rg}(\Gamma, F)$ is finite, we may assume that $\mathrm{cost}\,\Phi$ is finite as well. Let $\varepsilon \in \mathbb{R}_{>0}$. It suffices to show that there exists a subgroup $\Lambda \in F$ with

$$\frac{d(\Lambda) - 1}{[\Gamma : \Lambda]} \leq \mathrm{cost}(\Phi) + \varepsilon - 1.$$

In order to find such a subgroup, we will first replace Φ by a finitary approximation: First of all, by decomposing the domains of the partial $\mathcal{R}$-automorphisms in Φ according to the group elements acting, we may assume that every element of Φ is of the form $\gamma_i \cdot\; : A_i \longrightarrow \gamma_i \cdot A_i$ for some measurable subset $A_i \subset X$ and some $\gamma_i \in \Gamma$, and that Φ is countable.

Enumerating these partial $\mathcal{R}$-automorphisms and approximating their domains with exponential accuracy by open supersets (which is possible because μ is regular [87, Theorem 17.10]), we may replace Φ with a graphing consisting of partial translations with *open* domains and cost less than $\mathrm{cost}\,\Phi + \varepsilon$.

By Lemma 4.3.7 below, because X is compact, the Γ-action on X is continuous, and $\{\pi_\Lambda^{-1}(Z) \mid \Lambda \in F,\ Z \subset \Gamma/\Lambda\}$ is a countable basis of the topology on X that is closed under finite unions (as F is closed under finite intersections), we can replace this graphing by a finite graphing $\Psi = (\gamma_i \cdot\; : A_i \longrightarrow \gamma_i \cdot A_i)_{i \in I}$ with $\mathrm{cost}\,\Psi \leq \mathrm{cost}\,\Phi + \varepsilon$ and such that for each $i \in I$ there exists a $\Lambda_i \in F$ and a $g_i \in \Gamma$ with $A_i = \pi_{\Lambda_i}^{-1}(g_i \cdot \Lambda_i)$. Considering the finite intersection $\Lambda := \bigcap_{i \in I} \Lambda_i \subset \Gamma$, which also lies in F (and decomposing the partial automorphisms once more), we may furthermore assume that $\Lambda_i = \Lambda$ holds for all $i \in I$.

We will now show that Λ has the desired property: To this end, we consider the directed Γ-labelled multi-graph Y with vertex set $V := \Gamma/\Lambda$ and the labelled directed edges

$$E := \big\{ (\gamma_i \cdot g_i \cdot \Lambda, g_i \cdot \Lambda, \gamma_i) \mid i \in I \big\} \subset V \times V \times \Gamma.$$

Then

$$\operatorname{cost} \Psi = \frac{1}{[\Gamma : \Lambda]} \cdot |E|$$

and it remains to establish a suitable lower bound for $|E|$ in terms of $d(\Lambda)$.

We choose $i \in I$; let $v := g_i \cdot \Lambda \in V$ and let $\pi_1(Y, v)$ be the combinatorial fundamental group of Y. Then the map $\varphi \colon \pi_1(Y, v) \longrightarrow \Gamma$, defined on v-based cycles by the corresponding product of the edge-labels (or their inverses, if traversed in the opposite direction), is a well-defined group homomorphism (check!). Because Ψ is a graphing of $\mathcal{R}$, we obtain that Y is connected (check!) and that $\operatorname{im} \varphi = \Lambda$ (Lemma 4.3.8 below). In particular,

$$d(\Lambda) \leq d\big(\pi_1(Y, v)\big).$$

The group $\pi_1(Y, v)$ (which is isomorphic to the usual fundamental group of the geometric realisation of Y) is free; a free basis can be obtained by choosing an undirected spanning tree T of Y, collapsing this spanning tree T, and then taking the loops given by the edges of Y *not* contained in T. Hence,

$$d\big(\pi_1(Y, v)\big) = |E| - \#\text{edges of } T = |E| - \big(|V| - 1\big)$$
$$= |E| - [\Gamma : \Lambda] + 1;$$

for this, alternatively, one could have used $\chi(Y)$. Therefore, we obtain

$$\operatorname{cost} \Phi + \varepsilon - 1 \geq \operatorname{cost} \Psi - 1 = \frac{|E|}{[\Gamma : \Lambda]} - 1 = \frac{d(\pi_1(Y, v)) - 1 + [\Gamma : \Lambda]}{[\Gamma : \Lambda]} - 1$$
$$\geq \frac{d(\Lambda) - 1}{[\Gamma : \Lambda]}. \qquad \square$$

Lemma 4.3.7. *Let $\Gamma \curvearrowright X$ be a free continuous action of a finitely generated group on a compact second countable topological space, let μ be a Γ-invariant Borel probability measure on X. Let Φ be a countable graphing of $\mathcal{R}_{\Gamma \curvearrowright X}$ consisting of partial translations on open subsets of X and let O be a countable basis of the topology on X. Then there exists a finite graphing Ψ of $\mathcal{R}_{\Gamma \curvearrowright X}$ consisting of partial translations on finite unions of elements of O with $\operatorname{cost} \Psi \leq \operatorname{cost} \Phi$.*

Proof. In this proof, it will be more convenient to work with subsets of the topological space $X \times \Gamma$ (where Γ carries the discrete topology) than with subfamilies of $[\![\mathcal{R}_{\Gamma \curvearrowright X}]\!]$. For subsets $U, V \subset X \times \Gamma$, $n \in \mathbb{N}$, we use the following notation:

$$U \cdot V := \big\{ (x, \eta \cdot \gamma) \mid x \in X, \gamma, \eta \in \Gamma, (x, \gamma) \in V, (\gamma \cdot x, \eta) \in U \big\}$$
$$U^{n+1} := U^n \cdot U \quad \text{and} \quad U^0 := X \times \{e\}$$

Let $S \subset \Gamma$ be a finite generating set of Γ and let $\widetilde{S} := X \times S \subset X \times \Gamma$, which is compact (because X is compact and S is finite). By hypothesis, we may write $\Phi = (\gamma_j \cdot : A_j \longrightarrow \gamma_j \cdot A_j)_{j \in \mathbb{N}}$ with $\gamma_j \in \Gamma$ and open subsets $A_j \subset X$. Moreover, we can write each A_j as an ascending union $A_j = \bigcup_{m \in \mathbb{N}} A_{j,m}$ of finite unions of elements of O. For $n, m \in \mathbb{N}$, we set

$$\widetilde{\Phi}_{n,m} := \bigcup_{j=0}^{n} A_{j,m} \times \{\gamma_j\} \cup \bigcup_{j=0}^{n} (\gamma_j \cdot A_{j,m}) \times \{\gamma_j^{-1}\}.$$

Because Φ is a graphing of $\mathcal{R}_{\Gamma \curvearrowright X}$, we obtain (check!)

$$\bigcup_{n,m \in \mathbb{N}} \bigcup_{k \in \mathbb{N}} \widetilde{\Phi}_{n,m}^{k} = \bigcup_{k \in \mathbb{N}} \bigcup_{n,m \in \mathbb{N}} \widetilde{\Phi}_{n,m}^{k} - X \times \Gamma.$$

Moreover, each $\widetilde{\Phi}_{n,m}^{k}$ is open and as $\widetilde{S}$ is compact and the $(\widetilde{\Phi}_{n,m})_{(n,m) \in \mathbb{N} \times \mathbb{N}}$ are nested, there exist $n, m \in \mathbb{N}$ with $\bigcup_{k \in \mathbb{N}} \widetilde{\Phi}_{n,m}^{k} \supset \widetilde{S}$. Because S is a generating set of Γ, we conclude that $(\gamma_j \cdot : A_{j,m} \longrightarrow \gamma_j \cdot A_{j,m})_{j \in \{0,\dots,n\}}$ is a graphing of $\mathcal{R}_{\Gamma \curvearrowright X}$. $\qquad\square$

Lemma 4.3.8. *Let $Y = (V, E)$ be the Γ-labelled graph constructed in the proof of Theorem 4.3.5, let $v \in V$, and let $\varphi \colon \pi_1(Y, v) \longrightarrow \Gamma$ be the labelling homomorphism. Then $\operatorname{im} \varphi = \Lambda$.*

Proof. We have $\operatorname{im} \varphi \subset \Lambda$: Let $c = (v = v_0, v_1, \dots, v_{n+1} = v)$ be a v-based cycle in Y. Then there exist $i_0, \dots, i_{n+1} \in I$ and $\varepsilon_1, \dots, \varepsilon_{n+1} \in \{-1, 1\}$ such that

$$v_j = g_{i_j} \cdot \Lambda \quad \text{and} \quad g_{i_j} \cdot \Lambda = \gamma_{i_{j+1}}^{\varepsilon_{j+1}} \cdot g_{i_{j+1}} \cdot \Lambda$$

for all $j \in \{0, \dots, n\}$. Then $\varphi([c]) = \gamma_{i_1}^{\varepsilon_1} \cdots \cdot \gamma_{i_{n+1}}^{\varepsilon_{n+1}}$ lies in Λ because Λ is normal in Γ and

$$\gamma_{i_1}^{\varepsilon_1} \cdots \cdot \gamma_{i_{n+1}}^{\varepsilon_{n+1}} \cdot g_{i_0} \cdot \Lambda = \gamma_{i_1}^{\varepsilon_1} \cdots \cdot \gamma_{i_{n+1}}^{\varepsilon_{n+1}} \cdot g_{i_{n+1}} \cdot \Lambda = \cdots = g_{i_0} \cdot \Lambda.$$

Conversely, let $\lambda \in \Lambda$ and let $g \in \Gamma$ with $g \cdot \Lambda = v$. Because Ψ is a graphing of $\mathcal{R}$, we can reach the element $(\lambda \cdot g)_{\Delta \in F} = \lambda \cdot e_0 \in X$ from $e_0 := (g)_{\Delta \in F}$ through Ψ. In other words, there exist $n \in \mathbb{N}$, $i_0, \dots, i_{n+1} \in I$, and $\varepsilon_1, \dots, \varepsilon_{n+1} \in \{-1, 1\}$ such that

$$\lambda = \gamma_{i_1}^{\varepsilon_1} \cdots \cdot \gamma_{i_{n+1}}^{\varepsilon_{n+1}}$$

and $g_{i_{n+1}} \cdot \Lambda = g \cdot \Lambda = v$ as well as

$$\forall_{j \in \{0,\dots,n\}} \quad \gamma_{i_{j+1}}^{\varepsilon_{j+1}} \cdots \cdot \gamma_{i_{n+1}}^{\varepsilon_{n+1}} \cdot (g)_{\Delta \in F} \in \pi_\Lambda^{-1}(g_{i_j} \cdot \Lambda).$$

Inductively, we obtain $\gamma_{i_{j+1}}^{\varepsilon_{j+1}} \cdot g_{i_{j+1}} \cdot \Lambda = g_{i_j} \cdot \Lambda$. Because $\lambda \in \Lambda$, we have $g_{i_0} \cdot \Lambda = \lambda \cdot g \cdot \Lambda = g \cdot \Lambda = v$. Therefore, the sequence $(g_{i_0} \cdot \Lambda, \dots, g_{i_{n+1}} \cdot \Lambda)$ is a v-based cycle c in Y and $\varphi([c]) = \gamma_{i_1}^{\varepsilon_1} \cdots \cdot \gamma_{i_{n+1}}^{\varepsilon_{n+1}} = \lambda$. $\qquad\square$

Example 4.3.9 (rank gradients of amalgamated free products and HNN-extensions). Let $\Gamma = \Lambda_1 *_A \Lambda_2$ be a residually finite finitely generated amalgamated free product over an amenable group A. If Γ_* is a residual chain of Γ, then

$$\mathrm{rg}(\Gamma, \Gamma_*) = \mathrm{rg}\big(\Lambda_1, (\Lambda_1 \cap \Gamma_n)_{n \in \mathbb{N}}\big) + \mathrm{rg}\big(\Lambda_2, (\Lambda_2 \cap \Gamma_n)_{n \in \mathbb{N}}\big) + \frac{1}{|A|}.$$

If A is finite, then this statement can be shown through Bass–Serre theory [6, 86]. If A is infinite, then Pappas [116] showed how to combine Theorem 4.3.5 with Gaboriau's computations of cost of free products [63] to compute the rank gradients of Γ.

Similarly, if $\Gamma = \Lambda *_A$ is a residually finite finitely generated HNN-extension over an amenable group A and if Γ_* is a residual chain of Γ, then [116]

$$\mathrm{rg}(\Gamma, \Gamma_*) = \mathrm{rg}\big(\Lambda, (\Lambda \cap \Gamma_n)_{n \in \mathbb{N}}\big) + \frac{1}{|A|}.$$

A related application of Theorem 4.3.5 is Corollary 4.3.16.

4.3.2 The cost estimate for the first L^2-Betti number

There is a dynamical version of the rank gradient estimate (Corollary 3.3.5):

Theorem 4.3.10 (cost estimate [64, Corollaire 3.23]). *Let Γ be a group of finite type. Then*

$$b_1^{(2)}(\Gamma) - b_0^{(2)}(\Gamma) \leq \mathrm{cost}\,\Gamma - 1.$$

In particular: If Γ is infinite, then $b_1^{(2)}(\Gamma) \leq \mathrm{cost}\,\Gamma - 1$ and $\mathrm{cost}\,\Gamma \geq 1$.

Gaboriau's original proof is in terms of simplicial complexes over measured equivalence relations and covers the analogous estimate for measured equivalence relations. In the following, we will translate the proof of the rank estimate $b_1(\Gamma) \leq d(\Gamma)$ via projective resolutions into the setting of cost:

Proof. Let $\Gamma \curvearrowright X$ be an essentially free standard action of Γ (if Γ is finite, we can take the translation action on Γ; if Γ is infinite, we can take the Bernoulli shift). It suffices to show that every graphing Φ of $\mathcal{R} := \mathcal{R}_{\Gamma \curvearrowright X}$ satisfies

$$b_1^{(2)}(\Gamma) - b_0^{(2)}(\Gamma) \leq \mathrm{cost}\,\Phi - 1.$$

For this, we will use Φ to construct a suitable partial projective resolution and apply a basic Morse inequality (Exercise 4.E.7).

We may assume that $\Phi = (\varphi_i = \lambda_i \cdot : A_i \longrightarrow B_i)_{i \in I}$ and that $\mathrm{cost}\,\Phi = \sum_{i \in I} \mu(A_i) < \infty$; moreover, we set $L := L^\infty(X) \rtimes \Gamma$ and we equip $L^\infty(X)$ with the left L-module structure given by $f[\gamma] \cdot f' := f \cdot (\gamma \cdot f')$ for all $f, f' \in L^\infty(X)$ and all $\gamma \in \Gamma$. Analogously to the classical proof of the rank estimate, we consider the left L-modules

$$P_0 := L \quad \text{and} \quad Q := \bigoplus_{i \in I} L \cdot \chi_{A_i}[e]$$

and the L-homomorphisms

$$\varepsilon \colon P_0 \longrightarrow L^\infty(X)$$
$$f[\gamma] \longmapsto f$$
$$\partial_1^Q \colon Q \longrightarrow P_0$$
$$x \cdot \chi_{A_i}[e] \longmapsto x \cdot \chi_{A_i}[e] \cdot \big(1[e] - 1[\lambda_i]\big) = x \cdot \chi_{A_i}[e] - x \cdot \chi_{A_i}[\lambda_i].$$

Then $\operatorname{im} \partial_1^Q \subset \ker \varepsilon$. Unfortunately, in general, it is *not* clear that $\operatorname{im} \partial_1^Q$ is all of $\ker \varepsilon$. Therefore, we extend Q by a correction term to construct the module in degree 1: Let $\delta \in \mathbb{R}_{>0}$ and let $(\gamma_n)_{n \in \mathbb{N}}$ be an enumeration of Γ. For $k, n \in \mathbb{N}$, we set

$$A(k,n) := \big\{ x \in X \mid \exists_{i_1,\dots,i_n \in I} \ \exists_{\varepsilon_1,\dots,\varepsilon_n \in \{-1,1\}} \ \ \gamma_k \cdot x = \varphi_{i_n}^{\varepsilon_n} \circ \cdots \circ \varphi_{i_1}^{\varepsilon_1}(x) \big\},$$

which is a measurable subset of X. Because Φ is a graphing of $\mathcal{R}$, we obtain for all $k \in \mathbb{N}$ that $\bigcup_{n \in \mathbb{N}} A(k,n) = X$. Hence, there exists an $n_k \in \mathbb{N}$ such that $C_k := X \setminus \bigcup_{n=0}^{n_k} A(k,n)$ satisfies $\mu(C_k) \leq \delta \cdot 1/2^{k+1}$. We then consider the left L-module

$$M := \bigoplus_{k \in \mathbb{N}} L \cdot \chi_{C_k}[e]$$

and the L-homomorphism

$$\partial_1^M \colon M \longrightarrow P_0$$
$$x \cdot \chi_{C_k}[e] \longmapsto x \cdot \chi_{C_k}[e] \cdot \big(1[e] - 1[\gamma_k]\big).$$

Finally, we set $P_1 := Q \oplus M$ and $\partial_1 := \partial_1^Q \oplus \partial_1^M \colon P_1 \longrightarrow P_0$. By construction, we have $\operatorname{im} \partial_1 \subset \ker \varepsilon$ and also $\ker \varepsilon \subset \operatorname{im} \partial_1$ (Lemma 4.3.11 below). Moreover, P_0 and P_1 are projective L-modules. In other words, we constructed a partial L-projective resolution of $L^\infty(X)$: $0 \longleftarrow L^\infty(X) \xleftarrow{\ \varepsilon\ } P_0 \xleftarrow{\ \partial_1\ } P_1$. Thus, the Morse inequality (Exercise 4.E.7), the additivity of $\dim_{N\mathcal{R}}$, and Lemma 4.3.12 below imply that

$$b_1^{(2)}(\Gamma) - b_0^{(2)}(\Gamma) \leq \dim_{N\mathcal{R}}(N\mathcal{R} \otimes_L P_1) - \dim_{N\mathcal{R}}(N\mathcal{R} \otimes_L P_0)$$
$$= \dim_{N\mathcal{R}} N\mathcal{R} - \dim_{N\mathcal{R}}(N\mathcal{R} \otimes_L Q) - \dim_{N\mathcal{R}}(N\mathcal{R} \otimes_L M)$$
$$= 1 - \sum_{i \in I} \mu(A_i) - \sum_{k \in \mathbb{N}} \mu(C_k)$$
$$\leq 1 - \cot \Phi - \delta.$$

Taking $\delta \longrightarrow 0$ gives the desired estimate. $\qquad\qquad\square$

Lemma 4.3.11. *In the situation of the proof of Theorem 4.3.10, we have the inclusion* $\ker \varepsilon \subset \operatorname{im} \partial_1$.

Proof. Let $f = \sum_{\gamma \in \Gamma} f_\gamma [\gamma] \in \ker \varepsilon$. Then

$$f = \sum_{\gamma \in \Gamma} f_\gamma [\gamma] - \varepsilon(f)[e] = \sum_{\gamma \in \Gamma} f_\gamma [e] \cdot (1[\gamma] - 1[e]).$$

In view of L-linearity, it therefore suffices to show that $1[\gamma] - 1[e] \in \operatorname{im} \partial_1$ for each $\gamma \in \Gamma$. Let $\gamma \in \Gamma$. Then $\gamma = \gamma_k$ for some $k \in \mathbb{N}$ and (in $L^\infty(X)$)

$$1 = \chi_{C_k} + \sum_{n=0}^{n_k} \chi_{B(k,n)},$$

where we set $B(k,n) := A(k,n) \setminus \bigcup_{m=0}^{n-1} A(k,m)$. Therefore, we obtain

$$1[\gamma] - 1[e] = \chi_{C_k}[\gamma_k] - \chi_{C_k}[e] + \sum_{n=0}^{n_k} \left(\chi_{B(k,n)}[\gamma_k] - \chi_{B(k,n)}[e] \right).$$

The first difference lies in $\operatorname{im} \partial_1^M$. Therefore, it suffices to show that all other summands lie in ∂_1^Q. This in turn follows inductively (over the decomposition of the action of γ_k into the λ_i on appropriate subsets) from the construction of the $B(k,n)$ and the following observations (check!): For all $i \in I$, all measurable subsets $A \subset X$ with $\lambda_i^{\mp 1} \cdot A \subset B(k, n-1)$, and all $\eta \in \Gamma$, we have

$$\chi_A[\lambda_i^{\pm 1} \cdot \eta] - \chi_A[e] = \chi_A[e] \cdot \left(1[\lambda_i^{\pm 1}] \cdot (1[\eta] - 1[e]) + 1[\lambda_i^{\pm 1}] - 1[e] \right)$$
$$= -1[\lambda_i^{\pm 1}] \cdot \chi_{\lambda_i^{\mp 1} \cdot A}[e] \cdot (1[\eta] - 1[e])$$
$$+ \chi_A[e] \cdot \left(1[\lambda_i^{\pm 1}] - 1[e] \right).$$

By induction, we may assume that the first summand lies in $\operatorname{im} \partial_1^Q$. Moreover, if $A \subset A_i$, we have

$$\chi_A[e] \cdot (1[\lambda_i] - 1[e]) = -\partial_1^Q \left(\chi_A[e] \cdot \chi_{A_i}[e] \right)$$

and if $\lambda_i \cdot A \subset A_i$, we have

$$\chi_A[e] \cdot (1[\lambda_i^{-1}] - 1[e]) = -1[\lambda_i^{-1}] \cdot \chi_{\lambda_i \cdot A}[e] \cdot (1[\lambda_i] - 1[e]);$$

both of these elements lie in $\operatorname{im} \partial_1^Q$.   □

Lemma 4.3.12. *In the situation of the proof of Theorem 4.3.10, let $A \subset X$ be a measurable subset. Then $P := L \cdot \chi_A[e]$ is a projective left L-module and*

$$\dim_{N\mathcal{R}}(N\mathcal{R} \otimes_L P) = \mu(A).$$

Proof. The element $\chi_A[e] \in L$ is idempotent. Therefore, P is a projective left L-module. For the same reason, $N\mathcal{R} \otimes_L P \cong_{N\mathcal{R}} N\mathcal{R} \cdot \chi_A[e]$ (where we implicitly use the canonical inclusion $L \hookrightarrow \mathbb{C}\mathcal{R} \subset N\mathcal{R}$) is also a projective left $N\mathcal{R}$-module. Therefore, we obtain

$$\dim_{N\mathcal{R}}(N\mathcal{R} \otimes_L P) = \mathrm{pdim}_{N\mathcal{R}}(N\mathcal{R} \cdot \chi_A[e]) = \mathrm{tr}_{\mathcal{R}} \chi_A[e] = \mu(A). \qquad \square$$

4.3.3 Fixed price

By construction, cost of countable group actions is an orbit equivalence invariant. However, in general, cost of groups is hard to compute and the dependence on the underlying dynamical system remains a mystery:

Outlook 4.3.13 (fixed price problem). The fixed price problem asks for the (in)dependence of cost on the action:

Let Γ be a countable group and let $\Gamma \curvearrowright X$, $\Gamma \curvearrowright Y$ be essentially free standard actions. Do we then have

$$\mathrm{cost}(\Gamma \curvearrowright X) = \mathrm{cost}(\Gamma \curvearrowright Y) \quad (?!)$$

This problem is wide open. In fact, for finitely presented residually finite infinite groups Γ, in all known examples, one has $b_1^{(2)}(\Gamma) = \mathrm{cost}\,\mathcal{R}_{\Gamma \curvearrowright X} - 1 = \mathrm{rg}(\Gamma, \Gamma_*)$ for all essentially free ergodic standard actions $\Gamma \curvearrowright X$ and all residual chains Γ_* of Γ; this includes infinite amenable groups, free groups, surface groups ... [63].

Abért and Nikolov proved that the following (bold) conjectures exclude each other [7]:

- The [stable] rank vs. Heegaard genus conjecture for orientable compact hyperbolic 3-manifolds (the Heegaard genus of M equals $d(\pi_1(M))$).

- The fixed price conjecture.

Remark 4.3.14 (inner amenability). A group is *inner amenable* if it admits a conjugation invariant mean that has no atoms. Examples of inner amenable groups include all infinite amenable groups and all groups that have infinite centre. Similarly to infinite amenable groups, inner amenable countable groups have fixed price 1 [133] and thus vanishing first L^2-Betti number [37, Corollary D][133]. However, in contrast with amenability, inner amenability is *not* an ME invariant [81, p. 4].

Remark 4.3.15 (property (T)). An antagonist of amenability is property (T) (Remark 5.1.10). Countable infinite groups with property (T) are known to have cost 1 [80]; however, it remains an open problem whether they are of fixed price or not.

We conclude this chapter with a simple application to rank gradients:

Corollary 4.3.16. *Let* Γ *be a finitely generated residually finite infinite group of fixed price. Then, for every residual chain* Γ_* *of* Γ*, we have*

$$\mathrm{rg}(\Gamma, \Gamma_*) = \mathrm{rg}\,\Gamma.$$

In other words: In this case, the rank gradient does not *depend on the chosen residual chain.*

Proof. Applying Theorem 4.3.5 (twice) and the fixed price hypothesis shows that
$$\mathrm{rg}(\Gamma, \Gamma_*) = \mathrm{cost}(\Gamma \curvearrowright \widehat{\Gamma_*}) - 1 = \mathrm{cost}(\Gamma \curvearrowright \widehat{\Gamma}) - 1 = \mathrm{rg}\,\Gamma. \qquad \square$$

4.E Exercises

Exercise 4.E.1 (characterisations of residual finiteness). Let Γ be a finitely generated group. Show that the following are equivalent:

1. The group Γ is residually finite (i.e., it admits a residual chain).

2. For each $g \in \Gamma \setminus \{e\}$, there exists a finite group F and a group homomorphism $\varphi \colon \Gamma \longrightarrow F$ with $\varphi(g) \neq e$.

3. The diagonal homomorphism $\Gamma \longrightarrow \widehat{\Gamma} = \varprojlim_{\Lambda \in \mathrm{NF}(\Gamma)} \Gamma/\Lambda$ into the profinite completion of Γ is injective.

4. The diagonal action of Γ on the profinite completion $\widehat{\Gamma}$ is free.

Exercise 4.E.2 ((stable) orbit equivalence via equivalence relations). Reformulate the notion of (stable) orbit equivalence of standard actions in terms of the orbit relations (without using the group actions directly) and prove Corollary 4.2.8.

Exercise 4.E.3 (non-orbit equivalence of groups). Let $m, n \in \mathbb{N}$ and let $r_1, \ldots, r_m, s_1, \ldots, s_n \in \mathbb{N}_{\geq 2}$. Prove the following: If $m \neq n$, then $\prod_{j=1}^m F_{r_j}$ and $\prod_{j=1}^k F_{s_j}$ are *not* orbit equivalent.
Hints. L^2-Betti numbers ... This is a very special case of an ME rigidity result of Monod and Shalom [109].

Exercise 4.E.4 (L^2-Betti numbers of amenable groups). Compute the L^2-Betti numbers of amenable groups of finite type, i.e., prove Corollary 4.2.10.

Exercise 4.E.5 (L^2-Betti numbers of topological groups). Compute the L^2-Betti numbers (in the sense of Corollary 4.2.11) of the following topological groups:

1. $\mathbb{R}^{2019}$

2. $\mathrm{PSL}(2, \mathbb{R})$

3. $\mathbb{R}^{2019} \times \mathrm{PSL}(2, \mathbb{R})$

4. $\mathrm{PSL}(2, \mathbb{R}) \times \mathrm{PSL}(2, \mathbb{R})$

5. the three-dimensional real Heisenberg group

$$\left\{ \begin{pmatrix} 1 & x & z \\ 0 & 1 & y \\ 0 & 0 & 1 \end{pmatrix} \ \middle| \ x, y, z \in \mathbb{R} \right\} \subset \mathrm{SL}(3, \mathbb{R}).$$

Exercise 4.E.6 (an arithmetic lattice). For which $k \in \mathbb{N}$ is $b_k^{(2)}(\mathrm{SL}(2, \mathbb{Z}[\sqrt{2}]))$ non-trivial?

Hints. The group $\mathrm{SL}(2, \mathbb{R}) \times \mathrm{SL}(2, \mathbb{R})$ might help.

Exercise 4.E.7 (a Morse inequality). Let $\Gamma \curvearrowright X$ be an essentially free standard action of a countable group Γ, let $\mathcal{R} := \mathcal{R}_{\Gamma \curvearrowright X}$, and let

$$0 \longleftarrow L^\infty(X) \stackrel{\varepsilon}{\longleftarrow} P_0 \stackrel{\partial_1}{\longleftarrow} P_1$$

be an exact sequence of (left) $L^\infty(X) \rtimes \Gamma$-modules, where P_0 and P_1 are projective over $L^\infty(X) \rtimes \Gamma$. Show that

$$b_1^{(2)}(\Gamma) - b_0^{(2)}(\Gamma) \leq \dim_{N\mathcal{R}}(N\mathcal{R} \otimes_{L^\infty(X) \rtimes \Gamma} P_1) - \dim_{N\mathcal{R}}(N\mathcal{R} \otimes_{L^\infty(X) \rtimes \Gamma} P_0).$$

Hints. By (the proof of) Theorem 4.2.5, for all $k \in \mathbb{N}$, we have $b_k^{(2)}(\Gamma) = b_k^{(2)}(\mathcal{R}) = \dim_{N\mathcal{R}} \mathrm{Tor}_k^{L^\infty(X) \rtimes \Gamma}(N\mathcal{R}, L^\infty(X))$. We can compute this Tor-term by extending the given exact sequence to a projective resolution. Finally, we apply suitable monotonicity and additivity properties of $\dim_{N\mathcal{R}}$.

Exercise 4.E.8 (cost of finite groups). Let Γ be a finite group. Compute $\mathrm{cost}(\Gamma)$ and compare this result with the cost estimate for the first L^2-Betti number (Theorem 4.3.10).

Exercise 4.E.9 (cost of $\mathbb{Z}$). Use the Rokhlin lemma [90, Lemma 4.77] to compute the cost of every essentially free standard $\mathbb{Z}$-action.

Exercise 4.E.10 (ME cocycles [61, Section 2]). Let (Ω, μ) be an ME coupling between the countable groups Γ and Λ and let $Y, X \subset \Omega$ be corresponding finite measure fundamental domains for Γ and Λ, respectively. We consider the map

$$\alpha \colon \Gamma \times X \longrightarrow \Lambda$$
$$(\gamma, x) \longmapsto \lambda \in \Lambda \text{ with } \lambda \cdot \gamma \cdot x \in X.$$

Moreover, we equip X with the Γ-action

$$\bullet \colon \Gamma \times X \longrightarrow X$$
$$(\gamma, x) \longmapsto x' \in X \text{ with } \Lambda \cdot x' = \Lambda \cdot \gamma \cdot x.$$

Strictly speaking, the maps α and $\bullet$ are only well-defined almost everywhere.

1. Show that α is a cocycle, i.e., that for all $\gamma_1, \gamma_2 \in \Gamma$ and almost all $x \in X$, we have
$$\alpha(\gamma_1 \cdot \gamma_2, x) = \alpha(\gamma_1, \gamma_2 \bullet x) \cdot \alpha(\gamma_2, y).$$

2. What happens if we replace X by a different Λ-fundamental domain?

5

Invariant random subgroups

We will now consider an approximation theorem for covolume-normalised Betti numbers of uniform lattices in semi-simple Lie groups. We will first explain the statement of the theorem and give two instructive examples.

We will then sketch how ergodic theory, in the incarnation of invariant random subgroups, helps to handle such homology gradients and outline the structure of the proof of the theorem.

Overview of this chapter.

5.1	Generalised approximation for lattices	60
5.2	Two instructive examples	62
5.3	Convergence via invariant random subgroups	64
5.E	Exercises	71

Running example. lattices in $\mathrm{SL}(n, \mathbb{R})$ and in $\mathrm{SO}(n, 1)$, respectively

© The Author(s), under exclusive license to Springer Nature Switzerland AG 2020

C. Löh, *Ergodic Theoretic Methods in Group Homology*,

SpringerBriefs in Mathematics, https://doi.org/10.1007/978-3-030-44220-0_6

5.1 Generalised approximation for lattices

In the classical approximation theorem (Theorem 3.1.3), we looked at the limit behaviour of normalised ordinary Betti numbers of the form

$$\frac{b_k(\Gamma_n)}{[\Gamma : \Gamma_n]} \longrightarrow ?$$

If $(\Gamma_n)_{n \in \mathbb{N}}$ is a residual chain in an ambient group Γ of finite type, then this sequence indeed converges and the limit is $b_k^{(2)}(\Gamma)$ (Theorem 3.1.3).

We will now consider a version of the approximation theorem, where the conditions on the sequence $(\Gamma_n)_{n \in \mathbb{N}}$ are substantially relaxed, provided that these groups lie as lattices in a joint ambient group G. In view of Corollary 4.2.11, one might expect that the limit in this case is $b_k^{(2)}(G)$:

Remark 5.1.1 (classical approximation for lattices). Let G be a second countable locally compact topological group (with a given Haar measure), let $\Gamma \subset G$ be a lattice of finite type, and let $(\Gamma_n)_{n \in \mathbb{N}}$ be a residual chain in Γ (provided it exists). Then, for all $k \in \mathbb{N}$, we have

$$\lim_{n \to \infty} \frac{b_k(\Gamma_n)}{\mathrm{vol}(\Gamma_n \setminus G)} = \lim_{n \to \infty} \frac{b_k(\Gamma_n)}{[\Gamma : \Gamma_n]} \cdot \frac{[\Gamma : \Gamma_n]}{\mathrm{vol}(\Gamma_n \setminus G)}$$

$$= \frac{b_k^{(2)}(\Gamma)}{\mathrm{vol}(\Gamma \setminus G)} \qquad \text{(approximation theorem; Theorem 3.1.3)}$$

$$= b_k^{(2)}(G). \qquad \text{(by definition of } b_k^{(2)}(G); \text{ Corollary 4.2.11)}$$

5.1.1 Statement of the approximation theorem

We now state a selection of the approximation results of the "seven samurai" Abért, Bergeron, Biringer, Gelander, Nikolov, Raimbault, Samet [2, 3].

The terminology used in these results will be explained in Chapter 5.1.2, the proofs of Theorem 5.1.3 and Theorem 5.1.4 will be sketched in Chapter 5.3. Moreover, basic notions for lattices are recalled in Appendix A.3.

Setup 5.1.2. Let G be a connected centre-free semi-simple Lie group without compact factors with a chosen Haar measure, let $K \subset G$ be a maximal compact subgroup, and let $X := G/K$ be the associated symmetric space.

Theorem 5.1.3 (BS-approximation for lattices [2, Corollary 1.4]). *In the situation of Setup 5.1.2, let $(\Gamma_n)_{n \in \mathbb{N}}$ be a uniformly discrete sequence of uniform lattices in G such that $(\Gamma_n \setminus X)_{n \in \mathbb{N}}$ BS-converges to X and let $k \in \mathbb{N}$. Then*

$$\lim_{n\to\infty} \frac{b_k(\Gamma_n)}{\mathrm{vol}(\Gamma_n \setminus G)} = b_k^{(2)}(G).$$

Theorem 5.1.4 (a sufficient condition for BS-convergence [2, Theorem 1.5]). *In the situation of Setup 5.1.2, let G have property (T) and $\mathrm{rk}_{\mathbb{R}}\, G \geq 2$. Moreover, let $(\Gamma_n)_{n\in\mathbb{N}}$ be a sequence of pairwise non-conjugate irreducible lattices in G. Then $(\Gamma_n \setminus X)_{n\in\mathbb{N}}$ BS-converges to X.*

Combining these two theorems gives the following approximation result for lattices:

Corollary 5.1.5 (an approximation theorem for uniformly discrete lattices [2, Corollary 1.6]). *In the situation of Setup 5.1.2, let G have property (T) and $\mathrm{rk}_{\mathbb{R}}\, G \geq 2$. Moreover, let $(\Gamma_n)_{n\in\mathbb{N}}$ be a uniformly discrete sequence of pairwise non-conjugate irreducible lattices in G. Then*

$$\lim_{n\to\infty} \frac{b_k(\Gamma_n)}{\mathrm{vol}(\Gamma_n \setminus G)} = b_k^{(2)}(G).$$

In order to use Theorem 5.1.3 or Corollary 5.1.5 in concrete situations, it is useful to know the values on the right-hand side, which have been computed through locally symmetric spaces by analytic means.

Remark 5.1.6 (L^2-Betti numbers of locally symmetric spaces/semi-simple Lie groups). In the situation of Setup 5.1.2, the L^2-Betti numbers of G can be computed as follows [105, Theorem 5.12] (this goes back to Borel [25]): Let $\mathrm{frk}\, G := \mathrm{rk}_{\mathbb{C}}(G) - \mathrm{rk}_{\mathbb{C}}(K)$ be the *fundamental rank of G*. Then:

- If $\mathrm{frk}\, G \neq 0$, then $b_k^{(2)}(G) = 0$ for all $k \in \mathbb{N}$.

- If $\mathrm{frk}\, G = 0$, then, for all $k \in \mathbb{N}$

$$b_k^{(2)}(G) = \begin{cases} 0 & \text{if } 2 \cdot k \neq \dim X \\ \neq 0 & \text{if } 2 \cdot k = \dim X. \end{cases}$$

In particular, all (closed) locally symmetric spaces satisfy the Singer conjecture (Outlook 2.2.9).

5.1.2 Terminology

Benjamini–Schramm convergence is a probabilistic geometric notion of convergence, which has its origin in graph theory [18].

Definition 5.1.7 (BS-convergence [2]). In the situation of Setup 5.1.2, let $(\Gamma_n)_{n\in\mathbb{N}}$ be a sequence of lattices in G. Then the X-orbifolds $(\Gamma_n \setminus X)_{n\in\mathbb{N}}$ *BS-converge (Benjamini–Schramm converge)* to X if for every $R \in \mathbb{R}_{>0}$, the probability that the R-ball around a random point in $\Gamma_n \setminus X$ is isometric to the R-ball in X tends to 1, i.e., if

$$\forall_{R \in \mathbb{R}_{>0}} \quad \lim_{n \to \infty} \frac{\operatorname{vol}(R\text{-thin part of } \Gamma_n \setminus X)}{\operatorname{vol}(\Gamma_n \setminus X)} = 0.$$

The R-*thin part* of a Riemannian manifold M is $\{x \in M \mid \operatorname{injrad}_M(x) < R\}$.

Definition 5.1.8 (uniform discreteness). In the situation of Setup 5.1.2, a family $(\Gamma_i)_{i \in I}$ of lattices in G is *uniformly discrete*, if there exists an open neighbourhood $U \subset G$ of $e \in G$ with the property that

$$\forall_{i \in I} \ \forall_{g \in G} \quad g \cdot \Gamma_i \cdot g^{-1} \cap U = \{e\}.$$

Example 5.1.9. In the situation of Setup 5.1.2, let $\Gamma \subset G$ be a uniform lattice and let $(\Gamma_n)_{n \in \mathbb{N}}$ be a family of finite index subgroups of Γ. Then (Exercise 5.E.1):

- The family $(\Gamma_n)_{n \in \mathbb{N}}$ is uniformly discrete in G.

- If $(\Gamma_n)_{n \in \mathbb{N}}$ is a residual chain of Γ, then $(\Gamma_n \setminus X)_{n \in \mathbb{N}}$ BS-converges to X.

Remark 5.1.10 (property (T)). A locally compact second countable topological group G has *(Kazhdan's) property (T)*, if the trivial representation of G is an isolated point in the unitary dual of G (with respect to the Fell topology); equivalently, such groups can also be characterised in terms of (almost) invariant vectors in unitary representations [15].

- If $n \in \mathbb{N}_{\geq 3}$, then $\mathrm{SL}(n, \mathbb{R})$ has property (T). More generally, every simple Lie group G with $\operatorname{rk}_\mathbb{R} G \geq 2$ has property (T).

- The group $\mathrm{SL}(2, \mathbb{R})$ does *not* have property (T).

- Property (T) is an antagonist of amenability: If a group both is amenable and has property (T), then it is compact. This fact is beautifully exploited in the proof of the Margulis normal subgroup theorem [108].

5.2 Two instructive examples

As in the original paper [2], we will discuss two examples for Corollary 5.1.5.

5.2.1 Lattices in $\mathrm{SL}(n, \mathbb{R})$

Example 5.2.1 ([2, Example 1.7]). Let $d \in \mathbb{N}_{\geq 3}$, let $\Gamma \subset \mathrm{SL}(d, \mathbb{R})$ be a uniform lattice, and let $(\Gamma_n)_{n \in \mathbb{N}}$ be a sequence of distinct, finite index subgroups of Γ. Moreover, let $k \in \mathbb{N}$. Then

$$\lim_{n \to \infty} \frac{b_k(\Gamma_n)}{[\Gamma : \Gamma_n]} = 0.$$

To apply Corollary 5.1.5, we verify that the hypotheses are satisfied:

- Because $d \geq 3$, we know that $\mathrm{rk}_{\mathbb{R}}\, \mathrm{SL}(d, \mathbb{R}) \geq 2$ and that $\mathrm{SL}(d, \mathbb{R})$ has property (T) (Remark 5.1.10).

- As $\mathrm{SL}(d, \mathbb{R})$ is sufficiently irreducible, its lattices are also irreducible.

- The family $(\Gamma_n)_{n \in \mathbb{N}}$ is uniformly discrete (Example 5.1.9).

- Moreover, the sequence $(\Gamma_n)_{n \in \mathbb{N}}$ can, for each conjugacy class, contain at most finitely many members (this can be seen from the covolumes).

Hence, we obtain (from Corollary 5.1.5 and the argument in Remark 5.1.1)

$$\lim_{n \to \infty} \frac{b_k(\Gamma_n)}{[\Gamma : \Gamma_n]} = b_k^{(2)}\big(\mathrm{SL}(d, \mathbb{R})\big) \cdot \mathrm{vol}\big(\Gamma \setminus \mathrm{SL}(d, \mathbb{R})\big).$$

Moreover, the fundamental rank of $\mathrm{SL}(d, \mathbb{R})$ for $n \geq 3$ is non-zero. Hence, the right-hand side is 0 (Remark 5.1.6).

5.2.2 Why doesn't it work in rank 1 ?!

Example 5.2.2 (generalised approximation fails in rank 1 [2, p. 716]). There exist closed connected hyperbolic manifolds M with $d := \dim M \geq 3$ and the following property (Exercise 5.E.2): There exists a surjective group homomorphism $\pi \colon \pi_1(M) \longrightarrow F_2$.

We now consider the uniform lattice $\Gamma := \pi_1(M)$ in $\mathrm{SO}(d, 1)$ and the following sequence of subgroups: For each $n \in \mathbb{N}_{\geq 1}$, let $\Lambda_n \subset F_2$ be a subgroup of index n (these exist), and let

$$\Gamma_n := \pi^{-1}(\Lambda_n) \subset \Gamma := \pi_1(M).$$

Then $[\Gamma : \Gamma_n] = [F_2 : \Lambda_n] = n$.

In this situation, we have (which shows that the conclusion of Corollary 5.1.5 does *not* hold in this rank 1-situation):

- For each $n \in \mathbb{N}_{\geq 1}$,

$$\begin{aligned} b_1(\Gamma_n) = \dim_{\mathbb{Q}} H_1(\Gamma_n; \mathbb{Q}) = \mathrm{rk}_{\mathbb{Z}}\, H_1(\Gamma_n; \mathbb{Z}) &= \mathrm{rk}_{\mathbb{Z}}(\Gamma_n)_{\mathsf{Ab}} \\ &\geq \mathrm{rk}_{\mathbb{Z}}(\Lambda_n)_{\mathsf{Ab}} = \text{rank of the free group } \Lambda_n \\ &= n \cdot (2 - 1) + 1 \qquad\qquad \text{(Nielsen–Schreier)} \end{aligned}$$

and so $\liminf_{n \to \infty} \frac{b_1(\Gamma_n)}{[\Gamma : \Gamma_n]} = \liminf_{n \to \infty} \frac{n+1}{n} \geq 1$.

- In contrast, $b_1^{(2)}(\mathrm{SO}(d, 1)) = 0$ (Remark 5.1.6 or Outlook 2.2.8).

5.3 Convergence via invariant random subgroups

To prove the sufficient condition for BS-convergence (Theorem 5.1.4) and
the convergence of Betti numbers in the presence of BS-convergence (Theorem 5.1.3), one can make use of invariant random subgroups.

We will recall some basic terminology for invariant random subgroups
(Chapter 5.3.1) and reinterpret BS-convergence in terms of invariant ran-
dom subgroups (Chapter 5.3.2). We then sketch proofs of Theorem 5.1.4 and
Theorem 5.1.3.

5.3.1 Invariant random subgroups

Invariant random subgroups are a probabilistic version of normal subgroups,
defined on the Borel space of closed subgroups.

Definition 5.3.1 (the space of subgroups). Let G be a locally compact second
countable topological group.

- We write $\mathrm{Sub}(G)$ for the set of all closed subgroups of G, endowed
  with the subspace topology of the space of closed subsets of G with the
  Chabauty topology.

- The *Chabauty topology* [35] on $\mathrm{Sub}(G)$ is the topology generated by the
  basis

$$\{O_1(C) \mid C \subset G \text{ is compact}\} \cup \{O_2(U) \mid U \subset G \text{ is open}\},$$

  where for all compact $C \subset G$ and all open $U \subset G$, we set

$$O_1(C) := \{H \in \mathrm{Sub}(G) \mid H \cap C = \emptyset\}$$
$$O_2(U) := \{H \in \mathrm{Sub}(G) \mid H \cap U \neq \emptyset\}.$$

Remark 5.3.2 (convergence in the Chabauty topology). Let G be a locally
compact second countable topological group, let $(H_n)_{n\in\mathbb{N}}$ be a sequence
in $\mathrm{Sub}(G)$, and let $H \in \mathrm{Sub}(G)$. Then, by definition, $(H_n)_{n\in\mathbb{N}}$ converges
to H with respect to the Chabauty topology if and only if the following
conditions both hold (check!):

- For every $x \in H$, there exists a sequence $(x_n)_{n\in\mathbb{N}}$ in G with $x_n \in H_n$
  for all $n \in \mathbb{N}$ and $\lim_{n\to\infty} x_n = x$ in G.

- If $(x_n)_{n\in\mathbb{N}}$ is a sequence in G with $x_n \in H_n$ for all $n \in \mathbb{N}$, then every
  accumulation point of $(x_n)_{n\in\mathbb{N}}$ lies in H.

Definition 5.3.3 (invariant random subgroup). Let G be a locally compact second countable topological group.

- An *invariant random subgroup* is a Borel probability measure on $\mathrm{Sub}(G)$ that is invariant under the conjugation action of G on $\mathrm{Sub}(G)$.

- Let $\mathrm{IRS}(G)$ be the space of all invariant random subgroups on G (with the weak topology).

Example 5.3.4 (invariant random subgroups). Let G be a locally compact second countable group.

- If N is a closed normal subgroup of G, then the Dirac measure δ_N concentrated in $N \in \mathrm{Sub}(G)$ is an invariant random subgroup on G.

- If $\Gamma \subset G$ is a lattice, then the push-forward of the normalised Haar measure on $\Gamma \setminus G$ via

$$\Gamma \setminus G \longrightarrow \mathrm{Sub}(G)$$
$$\Gamma \cdot g \longmapsto g^{-1} \cdot \Gamma \cdot g$$

  is an invariant random subgroup on G (check!), which we will denote by μ_Γ. In particular, each sequence of lattices gives rise to a sequence of corresponding invariant random subgroups.

- If $G \curvearrowright (X, \nu)$ is an action of G on a standard Borel probability space by measure preserving Borel automorphisms, then the push-forward of ν under the stabiliser map

$$X \longrightarrow \mathrm{Sub}(G)$$
$$x \longmapsto G_x$$

  is an invariant random subgroup on G; the stabiliser subgroups are almost everywhere closed by a result of Varadarajan [138, Corollary 2.1.20]. Conversely, every invariant random subgroup on G arises in this way [5][2, Theorem 2.6].

Moreover, many exotic examples of invariant random subgroups exist [3, 27, 1]. However, we will focus on invariant random subgroups induced by normal subgroups or lattices.

Remark 5.3.5 (roots of invariant random subgroups). The idea of invariant random subgroups seems to be already implicitly present in Zimmer's work, in particular, in the Stück–Zimmer theorem. The name "invariant random subgroup" was first introduced by Abért, Glasner, and Virág [5], but similar notions appeared at the same time in the work of Vershik [134] and Bowen [26].

In the form of stabilisers of probability measure preserving actions, invariant random subgroups also occured in the work of Bergeron and Gaboriau on the first L^2-Betti number [19]. Moreover, in the graph-theoretic setting, invariant random subgroups of discrete countable groups correspond to unimodular random networks of Schreier graphs, as investigated by Aldous and Lyons [10]. More details on the history of invariant random subgroups are explained by Gelander [67, Section 10].

We will apply invariant random subgroups in order to prove approximation results for normalised Betti numbers. Another prominent application of invariant random subgroups is that they provide a means to reorganise and generalise rigidity results for lattices [95, 69, 70, 68].

5.3.2 Benjamini–Schramm convergence

In our setting, Benjamini–Schramm convergence can be translated into weak convergence of the corresponding invariant random subgroups:

Theorem 5.3.6 (BS-convergence and IRS-convergence [2, Corollary 3.8]). *In the situation of Setup 5.1.2, let $(\Gamma_n)_{n\in\mathbb{N}}$ be a sequence of lattices in G. Then the following are equivalent:*

1. *The sequence $(\Gamma_n \setminus X)_{n\in\mathbb{N}}$ BS-converges to X.*

2. *The sequence $(\mu_{\Gamma_n})_{n\in\mathbb{N}}$ of invariant random subgroups on G weakly converges to δ_1, the Dirac measure concentrated on the trivial subgroup 1.*

Proof. We show that each of theses conditions is equivalent to

3. For all $R \in \mathbb{R}_{>0}$, we have

$$\lim_{n\to\infty} \mu_{\Gamma_n}\big(\{H \in \mathrm{Sub}(G) \mid \mathrm{injrad}_{H\setminus X}(H \cdot e \cdot K) \leq R\}\big) = 0.$$

For $n \in \mathbb{N}$, we will use the following notation: Let $P_n := \mathrm{vol}(\,\cdot\,)/\mathrm{vol}(\Gamma_n \setminus X)$ be the normalised Riemannian measure on $\Gamma_n \setminus X$ and let μ_n be the G-invariant probability measure on $\Gamma_n \setminus G$ induced by the Haar measure on G. Then we can rewrite our measures as

$$\mu_{\Gamma_n} = s_{n*}\mu_n \quad \text{and} \quad P_n = p_{n*}\mu_n,$$

where $p_n \colon \Gamma_n \setminus G \longrightarrow \Gamma_n \setminus X$ is the canonical projection and $s_n \colon \Gamma_n \setminus G \longrightarrow \mathrm{Sub}(G)$ denotes the stabiliser map $\Gamma_n \cdot g \longmapsto g^{-1} \cdot \Gamma_n \cdot g$. Therefore, for all $R \in \mathbb{R}_{>0}$, we obtain

$$
\begin{aligned}
&P_n\big(\{x \in \Gamma_n \setminus X \mid \mathrm{injrad}_{\Gamma_n\setminus X}(x) \leq R\}\big)\\
&= \mu_n\big(\{\Gamma_n \cdot g \in \Gamma_n \setminus G \mid \mathrm{injrad}_{\Gamma_n\setminus X}(\Gamma_n \cdot g \cdot K) \leq R\}\big)\\
&= \mu_{\Gamma_n}\big(\{g^{-1} \cdot \Gamma_n \cdot g \in \mathrm{Sub}(G) \mid \mathrm{injrad}_{g^{-1}\Gamma_n g\setminus X}(g^{-1} \cdot \Gamma_n \cdot g \cdot e \cdot K) \leq R\}\big)\\
&= \mu_{\Gamma_n}\big(\{H \in \mathrm{Sub}(G) \mid \mathrm{injrad}_{H\setminus X}(H \cdot e \cdot K) \leq R\}\big).
\end{aligned}
$$

Now the definition of BS-convergence shows that the first and the third property are equivalent.

It remains to show that the second and the third property are equivalent: By the portmanteau theorem, we have that $(\mu_{\Gamma_n})_{n \in \mathbb{N}}$ weakly converges to δ_1 if and only if we have for all open subsets $U \subset \mathrm{Sub}(G)$ that

$$\liminf_{n \to \infty} \mu_{\Gamma_n}(U) \geq \delta_1(U).$$

By Lemma 5.3.7, this is equivalent to the third property.   □

Lemma 5.3.7 (more on the topology of $\mathrm{Sub}(G)$). *In the situation of Setup 5.1.2, the set $\{U_R \mid R \in \mathbb{R}_{>0}\}$ with*

$$U_R := \big\{ H \in \mathrm{Sub}(G) \mid \nexists_{\gamma \in H \setminus \{e\}} \ d_X(e \cdot K, \gamma \cdot K) \leq R \big\}$$

is a basis of open sets around the trivial subgroup 1 in $\mathrm{Sub}(G)$.

Proof. By definition, for all $R \in \mathbb{R}_{>0}$, we have $1 \in U_R$.

Moreover, U_R is open: We show that $\mathrm{Sub}(G) \setminus U_R$ is closed. In view of Remark 5.3.2, we take a sequence $(H_n)_{n \in \mathbb{N}}$ in $\mathrm{Sub}(G) \setminus U_R$ that converges to some $H \in \mathrm{Sub}(G)$ and show that $H \notin U_R$.

Because $H_n \notin U_R$, there exists a $\gamma_n \in H_n \setminus \{e\}$ with $d_X(e \cdot K, \gamma_n \cdot K) \leq R$. Without loss of generality, we may assume that there exists an open neighbourhood $U \subset G$ of e with

$$\forall_{n \in \mathbb{N}} \ \gamma_n \notin U.$$

This can be seen as follows: The set $D := \{g \in G \mid d_X(e \cdot K, g \cdot K) \leq R\}$ is a compact neighbourhood of e in G. Therefore, via the exponential map of G, we can find an open neighbourhood U of e in G with $U \subset D$ and

$$\forall_{g \in U \setminus \{e\}} \ \exists_{k \in \mathbb{N}} \ g^k \in D \setminus U.$$

Passing to such powers shows that we may assume without loss of generality that $\gamma_n \notin U$ for all $n \in \mathbb{N}$.

Because D is compact, $(\gamma_n)_{n \in \mathbb{N}}$ has an accumulation point $\gamma \in D$. As U is open, we know that $\gamma \notin U$; in particular, $\gamma \neq e$. Then, by definition of the topology on $\mathrm{Sub}(G)$, we have $\gamma \in H$ and γ witnesses that $H \notin U_R$.

It remains to show that every open neighbourhood of 1 in $\mathrm{Sub}(G)$ contains a U_R for a suitable $R \in \mathbb{R}_{>0}$: By definition of the Chabauty topology, we only need to consider sets of the form $O_1(C)$ and $O_2(U)$ for compact sets $C \subset G$ with $1 \notin K$ and open sets $U \subset G$ with $1 \in U$, respectively.

- If $U \subset G$ is open and $1 \in U$, then $O_2(U) = \mathrm{Sub}(G)$ and so every choice of $R \in \mathbb{R}_{>0}$ works.

- If $C \subset G$ is compact and $1 \notin C$, then we can just take R bigger than the finite number $\sup_{\gamma \in C} d_X(e \cdot K, \gamma \cdot K)$.   □

5.3.3   The accumulation point

We now prove Theorem 5.1.4. In view of Theorem 5.3.6, it suffices to establish the following:

Theorem 5.3.8 (the IRS accumulation point [2, Theorem 4.4]). *Let G be a connected centre-free semi-simple Lie group with property (T) and $\mathrm{rk}_{\mathbb{R}}\, G \geq 2$. Then the set*

$$\{\mu_\Gamma \mid \Gamma \text{ is an irreducible lattice in } G\}$$

has exactly one accumulation point in $\mathrm{IRS}(G)$, *namely* δ_1.

Sketch of proof. Clearly, δ_1 is an accumulation point of this set: The group G admits a lattice Γ. In particular, Γ is residually finite and thus contains a residual chain Γ_*. Then, $(\mu_{\Gamma_n})_{n\in\mathbb{N}}$ converges to δ_1, as can be seen through BS-convergence of the associated locally symmetric spaces (Exercise 5.E.1) and Theorem 5.3.6.

Why is δ_1 the only accumulation point? Using the Nevo–Stück–Zimmer theorem [130, 95], which is formulated in terms of actions and stabilisers, one can derive the following restriction on invariant random subgroups in higher rank [2, Theorem 4.1]:

- If G is a connected centre-free semi-simple Lie group with $\mathrm{rk}_{\mathbb{R}}\, G \geq 2$ and property (T), then every non-atomic irreducible invariant random subgroup of G is of the form μ_Γ for some irreducible lattice Γ in G.

  Here an invariant random subgroup of G is *irreducible*, if it is ergodic with respect to the action of every simple factor of G.

Let $(\Gamma_n)_{n\in\mathbb{N}}$ be a sequence of distinct irreducible lattices in G with the property that $(\mu_{\Gamma_n})_{n\in\mathbb{N}}$ converges in $\mathrm{IRS}(G)$, say to μ_∞. We show that $\mu_\infty = \delta_1$:

Because the Γ_n are irreducible, the simple factors of G act ergodically on $\Gamma_n \backslash G$ (by the Moore ergodicity theorem [16, Theorem III.2.1]), whence on μ_{Γ_n}. By the Glasner–Weiss theorem [71], then μ_∞ is also ergodic with respect to the simple factors of G. By the above version of Nevo–Stück–Zimmer theorem, we are reduced to the following cases:

- Lattice case: $\mu_\infty = \mu_\Lambda$ for an irreducedible lattice $\Lambda \subset G$ or

- atomic case: $\mu_\infty = \delta_N$ for a closed normal subgroup $N \subset G$.

The *lattice case* can be ruled out by Leuzinger's uniform lower bound of the first non-zero eigenvalue of the Laplacian on $\Gamma_n \backslash X$, a volume estimate, and Wang's finiteness theorem [2, p. 737f].

Regarding the *atomic case, assume* for a contradiction that $\mu_\infty = \delta_N$ for some normal subgroup $N \subset G$ of non-zero dimension. Then one can show that there exists a neighbourhood U of N in $\mathrm{Sub}(G)$ that contains no lattice. By the portmanteau theorem, we therefore obtain

$$0 = \liminf_{n \to \infty} \mu_{\Gamma_n}(U) \geq \mu_\infty(U) = \delta_N(U) = 1,$$

which is a contradiction: Therefore, the only remaining option is $\mu_\infty = \delta_N$, where N is a discrete normal subgroup of G. As G is connected, N is central, and as G is centre-free, N must be trivial; thus, $\mu_\infty = \delta_N = \delta_1$. □

This completes the proof of Theorem 5.1.4.

5.3.4 Reduction to Plancherel measures

We now turn to the proof of the Betti number approximation theorem, Theorem 5.1.3. As in the proof of the classical approximation theorem, the proof is based on a convergence of measures, the corresponding Plancherel measures:

In the situation of Theorem 5.1.3, we introduce the following notation:

- Let ν be the Plancherel measure on the unitary dual $\widehat{G}$ of G.

- For $n \in \mathbb{N}$, let ν_n be the relative Plancherel measure on $\widehat{G}$ of Γ_n in G, i.e.,

$$\nu_n := \frac{1}{\mathrm{vol}(\Gamma_n \setminus G)} \cdot \sum_{\pi \in \widehat{G}} m(\pi, \Gamma_n) \cdot \delta_\pi,$$

where $m(\pi, \Gamma_n)$ denotes the multiplicity of π in the right regular representation $L^2(\Gamma_n \setminus G)$ of G.

Similarly to the case of the classical approximation theorem (or the approximation theorem of DeGeorge–Wallach [42]), one can now express L^2-Betti numbers and covolume-normalised Betti numbers in terms of such measures, using eigenspace representations of the geometric Laplacian on (locally) symmetric spaces [2, Section 6.23] (this geometric approach leads to the same L^2-Betti numbers for the ambient group G [119, 120]): For the contributions of the Laplacian on k-forms, we obtain

$$b_k^{(2)}(G) = \nu^k(\{0\}) \quad \text{and} \quad \frac{b_k(\Gamma_n)}{\mathrm{vol}(\Gamma_n \setminus G)} = \nu_n^k(\{0\})$$

for all $n \in \mathbb{N}$, where the superscript k denotes a suitable push-forward measure.

Therefore, it suffices to prove a corresponding convergence result for the Plancherel measures.

5.3.5 Convergence of Plancherel measures

Theorem 5.3.9 (convergence of Plancherel measures [2, Theorem 6.7]). *In the situation of Setup 5.1.2, let $(\Gamma_n)_{n \in \mathbb{N}}$ be a uniformly discrete sequence of lat-*

tices in G such that $(\Gamma_n \backslash X)_{n \in \mathbb{N}}$ BS-converges to X. Then, for every relatively compact ν-regular open subset $A \subset \widehat{G}$, we have

$$\lim_{n \to \infty} \nu_n(A) = \nu(A).$$

The same conclusion also holds for relatively compact ν-regular open subsets of the tempered unitary dual of G.

Sketch of proof. Instead of working directly on the (complicated) space of all Borel measures on the unitary dual $\widehat{G}$, one shows that the (relative) Plancherel measures in question are reflected through continuous linear forms (given by integration) on a space $\mathcal{F}(\widehat{G})$ of sufficiently continuous bounded functions on $\widehat{G}$.

Because the locally symmetric spaces BS-converge to X, the corresponding invariant random subgroups $(\mu_{\Gamma_n})_{n \in \mathbb{N}}$ weakly converge to δ_1 (Theorem 5.3.6). The convergence of the invariant random subgroups, uniform discreteness, and the Plancherel formula by Harish–Chandra then show that the integration functionals on $\mathcal{F}(\widehat{G})$ of the $(\Gamma_n)_{n \in \mathbb{N}}$ converge to that of G.

Then Sauvageot's density principle can be applied to conclude that the Plancherel measures converge as stated [2, Section 6]. □

Corollary 5.3.10 (pointwise convergence of Plancherel measures [2, Corollary 6.9]). *In the situation of Setup 5.1.2, let $(\Gamma_n)_{n \in \mathbb{N}}$ be a uniformly discrete sequence of lattices in G such that $(\Gamma_n \backslash X)_{n \in \mathbb{N}}$ BS-converges to X. Then, for every $\pi \in \widehat{G}$, we have*

$$\lim_{n \to \infty} \nu_n(\{\pi\}) = \nu(\{\pi\}).$$

Proof. We distinguish two cases:

- If $\nu(\{\pi\}) = 0$, we take a sequence $(A_k)_{k \in \mathbb{N}}$ of relatively compact ν-regular open subsets of $\widehat{G}$ with $\{\pi\} = \bigcap_{k \in \mathbb{N}} A_k$. Then the convergence in Theorem 5.3.9 shows that

$$0 \leq \lim_{n \to \infty} \nu_n(\{\pi\}) \leq \limsup_{n \to \infty} \nu_n\left(\bigcap_{k \in \mathbb{N}} A_k\right) \leq \nu\left(\bigcap_{k \in \mathbb{N}} A_k\right) = \nu(\{\pi\}) = 0.$$

- If $\nu(\{\pi\}) \neq 0$, then π is a discrete series representation of G and thus an isolated point in the tempered unitary dual of G. Therefore, we can apply Theorem 5.3.9 directly to $\{\pi\}$. □

Alternatively, one can also give a more direct proof of this corollary [2, Section 6.10].

This finishes the proof outline of Theorem 5.1.3.

The BS-convergence results for Betti number gradients have been extended and generalised in many ways [3, 58, 49, 4, 22, 27, 111, 69, 1, 127, 92, 43].

5.E Exercises

Exercise 5.E.1 (BS-convergence, uniform discreteness). In the situation of Setup 5.1.2, let $\Gamma \subset G$ be a uniform lattice and let $(\Gamma_n)_{n\in\mathbb{N}}$ be a family of finite index subgroups of Γ. Show the following:

1. There exists an open neighbourhood U of e in G with

$$\forall_{g\in G} \quad g \cdot \Gamma \cdot g^{-1} \cap U = \{e\}.$$

2. The family $(\Gamma_n)_{n\in\mathbb{N}}$ is uniformly discrete in G.

3. If $(\Gamma_n)_{n\in\mathbb{N}}$ is a residual chain of Γ, then $(\Gamma_n \backslash X)_{n\in\mathbb{N}}$ BS-converges to X.
   Hints. Given $R \in \mathbb{R}_{>0}$, what happens with the R-thin part of $\Gamma_n \backslash X$ for large n ?

Exercise 5.E.2 (fundamental groups of hyperbolic manifolds). Why are there closed hyperbolic manifolds of dimension at least 3, whose fundamental group surjects onto the free group of rank 2 ?
Hints. You will need some (non-trivial) tool to do this, e.g., virtual properties of hyperbolic 3-manifolds [9].

Exercise 5.E.3 (lattices in rank 1). In Example 5.2.2 (generalised approximation fails in rank 1), (why) is it important to work in dimension at least 3 ?

Exercise 5.E.4 (isolated subgroups [66, Exercise 3]).

1. Let Γ be a discrete group. Show that Γ is an isolated point of $\mathrm{Sub}(\Gamma)$ if and only if Γ is finitely generated.

2. Show that S^1 (with the standard topology) is *not* an isolated point of $\mathrm{Sub}(S^1)$.

Exercise 5.E.5 (convergence of measures?). Let $(\mu_n)_{n\in\mathbb{N}}$ be a sequence of Borel probability measures on $[0,1]$ and let μ be a Borel probability measure on $[0,1]$ with the property that

$$\lim_{n\to\infty} \mu_n(U) = \mu(U)$$

holds for all open subsets $U \subset [0,1]$. Moreover, let $(U_k)_{k\in\mathbb{N}}$ be a nested decreasing sequence of open subsets of $[0,1]$ and let $V := \bigcap_{k\in\mathbb{N}} U_k$.

1. Do we always have $\limsup_{n\to\infty} \mu_n(V) \leq \mu(V)$?

2. Do we always have $\liminf_{n\to\infty} \mu_n(V) \geq \mu(V)$?

6

Simplicial volume

In the past chapters, we explored the interaction between different views on L^2-Betti numbers. We will now consider a similar interaction for simplicial volume.

Simplicial volume is a numerical topological invariant of manifolds, measuring the "size" of manifolds in terms of the "number" of singular simplices. Simplicial volume is also related to Riemannian volume and geometric structures on manifolds and therefore is a suitable invariant for certain geometric rigidity phenomena.

We quickly survey basic properties of simplicial volume and its similarities/differences with L^2-Betti numbers and related invariants. In particular, we will discuss the residually finite and the dynamical view on simplicial volume. After this survey, we will sketch some of the prototypical arguments for simplicial volume that involve ergodic theory.

Overview of this chapter.

6.1	Simplicial volume	74
6.2	The residually finite view	75
6.3	The dynamical view	77
6.4	Basic proof techniques	79
6.E	Exercises	92

Running example. hyperbolic manifolds, tori

© The Author(s), under exclusive license to Springer Nature Switzerland AG 2020
C. Löh, *Ergodic Theoretic Methods in Group Homology*,
SpringerBriefs in Mathematics, https://doi.org/10.1007/978-3-030-44220-0_7

6.1 Simplicial volume

Simplicial volume was first introduced by Gromov in his proof of Mostow rigidity [110, 73]. Simplicial volume is the ℓ^1-semi-norm of the $\mathbb{R}$-fundamental class, defined in terms of *singular* chains:

Definition 6.1.1 (simplicial volume). Let M be an oriented closed connected n-manifold.

- If R is a commutative ring with unit, then $Z(M; R) \subset C_n^{\mathrm{sing}}(M; R)$ denotes the set of all singular R-fundamental cycles of M.

- The *simplicial volume of M* is defined by

$$\|M\| := \inf\{|c|_1 \mid c \in Z(M; \mathbb{R})\} \in \mathbb{R}_{\geq 0},$$

where $|\sum_{\sigma \in \mathrm{map}(\Delta^n, M)} a_\sigma \cdot \sigma|_1 := \sum_{\sigma \in \mathrm{map}(\Delta^n, M)} |a_\sigma|$.

The main tools to compute simplicial volume are concrete geometric constructions on the singular chain complex and bounded cohomology. By now, simplicial volume has been computed in a rich class of examples [73, 137, 94, 31, 76, 39, 33, 32, 97, 77, 78]. For simplicity, we will only consider a selection of properties that fit in the context of L^2-Betti numbers and related invariants:

- **Multiplicativity.** If M is an oriented closed connected manifold and $N \longrightarrow M$ is a d-sheeted covering, then [73] (Exercise 6.E.1)

$$\|N\| = d \cdot \|M\|.$$

- **Hyperbolicity.** If M is an oriented closed connected hyperbolic n-manifold, then [73, 132, 17]

$$\|M\| = \frac{\mathrm{vol}(M)}{\text{volume of the ideal regular } n\text{-simplex in } \mathbb{H}^n}.$$

- **Amenability.** If M is an oriented closed connected manifold of non-zero dimension with amenable fundamental group, then [73, 82]

$$\|M\| = 0.$$

- **Proportionality principle.** If M and N are oriented closed connected Riemannian manifolds with isometric Riemannian universal covering, then [73, 132, 129]

$$\frac{\|M\|}{\mathrm{vol}(M)} = \frac{\|N\|}{\mathrm{vol}(N)}.$$

- **Low dimensions.** Simplicial volume is known in dimension ≤ 3 [73, 128].

In view of these striking similarities with L^2-Betti numbers, Gromov asked the following question.

Question 6.1.2 ([74, p. 232]). Let M be an oriented closed connected aspherical manifold with $\|M\| = 0$.

- Does this imply that $\chi(M) = 0$?!

- Does this imply that $b_n^{(2)}(M) = 0$ for all $n \in \mathbb{N}$?!

In general, these questions are wide open. Betti number estimates become available when one passes to a more integral setting, as in the residually finite view (Chapter 6.2) or the dynamical view (Chapter 6.3).

6.2 The residually finite view

Taking integral instead of real coefficients leads to integral simplicial volume. In the residually finite view, we then stabilise along finite-sheeted coverings:

Definition 6.2.1 (stable integral simplicial volume). Let M be an oriented closed connected n-manifold.

- The *integral simplicial volume of M* is defined by

$$\|M\|_{\mathbb{Z}} := \inf\{|c|_1 \mid c \in Z(M;\mathbb{Z})\} \in \mathbb{N}.$$

- The *stable integral simplicial volume of M* is defined by

$$\|M\|_{\mathbb{Z}}^{\infty} := \inf_{(p:\, N \to M) \in \mathrm{F}(M)} \frac{\|N\|_{\mathbb{Z}}}{|\deg(p)|} \in \mathbb{R}_{\geq 0},$$

where $\mathrm{F}(M)$ denotes the class of all finite-sheeted coverings of M.

Stable integral simplicial volume has the following properties:

- **Multiplicativity.** If M is an oriented closed connected manifold and $N \longrightarrow M$ is a d-sheeted covering, then a straightforward transfer argument (Exercise 6.E.1) shows that

$$\|N\|_{\mathbb{Z}}^{\infty} = d \cdot \|M\|_{\mathbb{Z}}^{\infty}.$$

- **Hyperbolicity.** Let M be an oriented closed connected hyperbolic n-manifold.

  - If $n = 2$, then $\|M\|_{\mathbb{Z}}^{\infty} = \|M\|$, by direct computation [73].

- If $n = 3$, then also $\|M\|_{\mathbb{Z}}^{\infty} = \|M\|$, by indirect computation [60]; indeed, the only known proof requires passage through the dynamical view and non-trivial results from ergodic theory! We will sketch this argument in Chapter 6.4.5.

- If $n \geq 4$, then the ratio $\|M\|_{\mathbb{Z}}^{\infty} / \|M\|$ is uniformly bounded away from 1 [59]. In particular, approximation fails in general for simplicial volume of aspherical manifolds.

- **Amenability.** Let M be an oriented closed connected aspherical manifold of non-zero dimension with amenable residually finite fundamental group. Then [60] (Chapter 6.4.4)

$$\|M\|_{\mathbb{Z}}^{\infty} = 0.$$

- **Betti number estimate.** If M is an oriented closed connected manifold, then the explicit description of Poincaré duality on the singular chain complex shows that, for all $k \in \mathbb{N}$, we have [75]

$$b_k(M) \leq \|M\|_{\mathbb{Z}};$$

we will explain this in Chapter 6.4.2. Stabilisation therefore leads to an L^2-Betti number bound. In addition, log-torsion-homology [gradient] bounds in terms of [stable] integral simplicial volume are also known [125].

- **Rank gradient estimate.** If M is an oriented closed connected manifold with residually finite (infinite) fundamental group, then [98] (Chapter 6.4.3)

$$\operatorname{rg}(\pi_1(M)) \leq \|M\|_{\mathbb{Z}}^{\infty}.$$

- **Low dimensions.** Stable integral simplicial volume of oriented closed connected *aspherical* manifolds coincides with ordinary simplicial volume up to dimension 3. For dimensions 1 and 2, this follows from the previous results; for dimension 3, again ergodic theory is helpful [55] (we will sketch the argument in Chapter 6.4.6).

In connection with Question 6.1.2 and the approximation results for Betti numbers, it is therefore natural to consider the following (open) problems:

Question 6.2.2 ([60, Question 1.12]). Let M be an oriented closed connected aspherical manifold with $\|M\| = 0$ and residually finite fundamental group. Does this imply that $\|M\|_{\mathbb{Z}}^{\infty} = 0$?!

Question 6.2.3. Let M be an oriented closed connected locally symmetric space of non-compact type of higher rank. Do we then have $\|M\| = \|M\|_{\mathbb{Z}}^{\infty}$?!

Question 6.2.4. For which (contractible) Riemannian universal covering manifolds does stable integral simplicial volume satisfy the proportionality principle?

Furthermore, as in the case of the rank gradient, it is not clear how/whether the choice of specific residual chains of subgroups/finite coverings affects the limit of the normalised integral simplicial volumes.

6.3 The dynamical view

In the dynamical view, we use twisted coefficients, based on standard actions of the fundamental group.

Definition 6.3.1 (integral foliated simplicial volume [126]). Let M be an oriented closed connected n-manifold with fundamental group Γ and let $\alpha = (\Gamma \curvearrowright (X, \mu))$ be a standard action.

- Let $Z(M; \alpha)$ be the set of all singular n-cycles of M in the twisted chain complex $C_*^{\text{sing}}(M; L^\infty(X, \mu; \mathbb{Z})) = L^\infty(X; \mathbb{Z}) \otimes_{\mathbb{Z}\Gamma} C_*^{\text{sing}}(\widetilde{M}; \mathbb{Z})$ that are homologous (in this twisted chain complex) to ordinary integral fundamental cycles of M.

- The α-*parametrised simplicial volume* of M is defined by

$$|M|^{\alpha} := \inf\left\{ \sum_{j=1}^{n} \int_X |f_j| \, d\mu \ \Big| \ \sum_{j=1}^{n} f_j \otimes \sigma_j \in Z(M; \alpha) \right\} \in \mathbb{R}_{\geq 0}.$$

- The *integral foliated simplicial volume* of M is then defined by

$$|M| := \inf_{\alpha \in S(\Gamma)} |M|^{\alpha},$$

where $S(\Gamma)$ denotes the class of all standard actions of Γ.

Remark 6.3.2 (comparing/combining the different views). These simplicial volumes are related as follows: For all oriented closed connected manifolds M, we have [100]

$$\|M\| \leq |M| \leq \|M\|_{\mathbb{Z}}^{\infty}$$

and (if the fundamental group Γ of M is residually finite) [60] (Chapter 6.4.1)

$$|M|^{\Gamma \curvearrowright \widehat{\Gamma}} = \|M\|_{\mathbb{Z}}^{\infty}.$$

This relation between the stable integral simplicial volume and the parametrised simplicial volume has recently proved useful to compute stable integral simplicial volume in cases where direct approaches failed, such as the case of hyperbolic/aspherical 3-manifolds [60, 55], manifolds with S^1-actions [52], or higher-dimensional graph-manifolds [53]. We will explain this in more detail in Chapter 6.4.

Integral foliated simplicial volume has the following properties:

- **Multiplicativity.** If M is an oriented closed connected manifold and $N \longrightarrow M$ is a d-sheeted covering, then a transfer argument [100] shows that

$$|N| = d \cdot |M|.$$

- **Hyperbolicity.** Let M be an oriented closed connected hyperbolic n-manifold.

  - If $n = 2$, then $|M| = \|M\|$, by direct computation [100].
  - If $n = 3$, then $|M| = \|M\|$, using an argument involving measure equivalence [100] (Chapter 6.4.5).
  - If $n \geq 4$, then $|M| / \|M\|$ is uniformly bounded away from 1 [60].

- **Amenability.** If M is an oriented closed connected aspherical manifold of non-zero dimension with amenable fundamental group, then [60] (Chapter 6.4.4)

$$|M| = 0.$$

- **L^2-Betti number estimate.** If M is an oriented closed connected manifold, then a parametrised Poincaré duality argument shows that, for all $k \in \mathbb{N}$, we have [126] (Chapter 6.4.2)

$$b_k^{(2)}(M) \leq |M|.$$

- **Cost estimate.** If M is an oriented closed connected manifold, then [99] (Chapter 6.4.3)

$$\mathrm{cost}\big(\pi_1(M)\big) - 1 \leq |M|.$$

- **Low dimensions.** Integral foliated simplicial volume of oriented closed connected *aspherical* manifolds coincides with ordinary simplicial volume up to dimension 3. For dimensions 1 and 2, this follows from the previous results; for dimension 3, geometrisation, the computation of the hyperbolic case, and amenable additivity is used [55] (Chapter 6.4.6).

As in the case of cost, it is not clear how/whether the choice of specific essentially free standard actions affects the corresponding parametrised simplicial volume [60, Section 1.5].

While simplicial volume and integral foliated simplicial volume are different for general aspherical manifolds (e.g., for hyperbolic manifolds in high dimensions), they might still have the same vanishing behaviour. In the context of Question 6.1.2 it is therefore natural ask the following:

Question 6.3.3 ([60, Question 1.12]). Let M be an oriented closed connected aspherical manifold with $\|M\| = 0$. Does this imply that $|M| = 0$?!

Question 6.3.4 ([99, Question 1.5]). Let M be an oriented closed connected aspherical manifold with $\|M\| = 0$. Does this imply that $\operatorname{cost} \pi_1(M) = 1$?!

Question 6.3.5. Let M and N be oriented closed connected aspherical manifolds with $|M| = 0$ and $\pi_1(M) \sim_{\mathrm{ME}} \pi_1(N)$. Does this imply that $|N| = 0$?! What about ordinary simplicial volume? [14, Question 1.14] What about quasi-isometry instead of measure equivalence?

Outlook 6.3.6 (further connections). In addition, the various simplicial volumes and L^2-Betti numbers also show similar behaviour with respect to:

- S^1-actions [105, 73, 137, 52] and circle foliations [105, 73, 34]

- graph manifolds [105, 53]

- amenable covers [124, 28, 73, 82, 101]

- (minimal) volume estimates [124, 28]

- (certain) mapping tori [105, 33]

6.4 Basic proof techniques

In the following, we collect (sketch) proofs of some of the above results for stable integral simplicial volume and integral foliated simplicial volume. We will focus on arguments that are related to $(L^2$-)Betti numbers or ergodic theory and explain the synergy between the residually finite and the dynamical view.

6.4.1 The role of the profinite completion

Similar to the rank gradient (Theorem 4.3.5), stable integral simplicial volume can also be expressed in terms of its ergodic theoretic companion:

Theorem 6.4.1 (parametrised simplicial volume of the profinite completion [100, Theorem 6.6, Remark 6.7]). *Let M be an oriented closed connected manifold with residually finite fundamental group Γ. Then*

$$\|M\|_{\mathbb{Z}}^\infty = |M|^{\Gamma \curvearrowright \widehat{\Gamma}}.$$

Proof. We begin with the estimate "$\geq$": Let $\Lambda \subset \Gamma$ be a finite index subgroup. Then the translation action $\Gamma \curvearrowright \Gamma/\Lambda$ and the normalised counting measure on Γ/Λ provide an example of a standard Γ-space. A straightforward computation shows that [100, Corollary 4.27]

$$|M|^{\Gamma \cap \widehat{\Gamma}} \leq |M|^{\Gamma \cap \Gamma/\Lambda} = \frac{1}{[\Gamma : \Lambda]} \cdot \|\Lambda \setminus \widetilde{M}\|_{\mathbb{Z}}.$$

Taking the infimum over all finite index subgroups of Γ (and the fundamental correspondence of covering theory) shows that $|M|^{\Gamma \cap \widehat{\Gamma}} \leq \|M\|_{\mathbb{Z}}^{\infty}$.

We now establish the converse estimate "$\leq$": Let $\widehat{c} \in Z(M; L^{\infty}(\widehat{\Gamma}, \mathbb{Z}))$ and let $\varepsilon \in \mathbb{R}_{>0}$. Then it suffices to show that $\|M\|_{\mathbb{Z}}^{\infty} \leq |\widehat{c}|_1 + (n+2) \cdot \varepsilon$, where $|\cdot|_1$ denotes the "norm" used in the definition of $|M|^{\Gamma \cap \widehat{\Gamma}}$. Because $\widehat{c}$ is a fundamental cycle, there exist $z \in Z(M; \mathbb{Z})$ and $\widehat{b} \in C_{n+1}^{\mathrm{sing}}(M; L^{\infty}(\widehat{\Gamma}, \mathbb{Z}))$ with

$$\widehat{c} = z + \partial \widehat{b}.$$

As in the proof for the rank gradient (Theorem 4.3.5), we will now use approximation by cylinder sets: We consider the $\mathbb{Z}\Gamma$-submodule

$$L := \mathrm{Span}_{\mathbb{Z}}\{\chi_{\pi_{\Lambda}^{-1}(A)} \mid \Lambda \in F(\Gamma), \ A \subset \Gamma/\Lambda\}$$

of $L^{\infty}(\widehat{\Gamma}, \mathbb{Z})$, where $\pi_{\Lambda} \colon \widehat{\Gamma} \longrightarrow \Gamma/\Lambda$ denotes the canonical projection. Approximating the supports of the finitely many steps of the coefficient functions of $\widehat{b}$ by cylinder sets (Lemma 6.4.2), we find $b \in C_{n+1}^{\mathrm{sing}}(M; L)$ with $|b - \widehat{b}|_1 \leq \varepsilon$. Then $c := z + \partial b$ is a cycle in $C_*^{\mathrm{sing}}(M; L)$ that is homologous to z (whence a fundamental cycle) and satisfies

$$|c|_1 \leq |z + \partial \widehat{b}|_1 + |\partial(b - \widehat{b})|_1 \leq |\widehat{c}|_1 + (n+2) \cdot \varepsilon.$$

Taking the (finite!) intersection Λ of the finite index subgroups appearing in the coefficient functions of c and b shows that (check!)

$$|c|_1 \geq |M|^{\Gamma \cap \Gamma/\Lambda} = \frac{1}{[\Gamma : \Lambda]} \cdot \|\Lambda \setminus \widetilde{M}\|_{\mathbb{Z}}.$$

Therefore, $\|M\|_{\mathbb{Z}}^{\infty} \leq |c|_1 \leq |\widehat{c}|_1 + (n+2) \cdot \varepsilon.$ $\square$

Lemma 6.4.2. *Let Γ be a finitely generated residually finite group, let $A \subset \widehat{\Gamma}$ be a measurable subset, and let $\varepsilon \in \mathbb{R}_{>0}$. Then there exists a finite index subgroup $\Lambda \subset \Gamma$ and a subset $Z \subset \Gamma/\Lambda$ with*

$$\mu\bigl(A \bigtriangleup \pi_{\Lambda}^{-1}(Z)\bigr) \leq \varepsilon,$$

where $\pi_{\Lambda} \colon \widehat{\Gamma} \longrightarrow \Gamma/\Lambda$ denotes the canonical projection.

Proof. The set $\{\pi_{\Lambda}^{-1}(Z) \mid \Lambda \in F(\Gamma), \ Z \subset \Gamma/\Lambda\}$ is a basis of the topology on $\widehat{\Gamma}$ and the probability measure μ on $\widehat{\Gamma}$ is regular [87, Theorem 17.10]. Hence, there exist sequences $(\Lambda_n)_{n \in \mathbb{N}}$ in $F(\Gamma)$ and $(Z_n \subset \Gamma/\Lambda_n)_{n \in \mathbb{N}}$ such that $B := \bigcup_{n \in \mathbb{N}} \pi_{\Lambda_n}^{-1}(Z_n)$ satisfies

$$A \subset B \quad \text{and} \quad \mu(B \setminus A) \leq \frac{\varepsilon}{2}.$$

Then there exists an $N \in \mathbb{N}$ such that the initial part $B_N := \bigcup_{n=0}^{N} \pi_{\Lambda_n}^{-1}(Z_n)$
satisfies $\mu(A \triangle B_N) \leq \varepsilon$. Because $\Lambda := \bigcap_{n=0}^{N} \Lambda_n$ has finite index in Γ, we can
rewrite B_N in the desired form.   □

More generally, similar arguments also work for all residual chains and the
associated profinite completions.

6.4.2 Betti number estimates

Betti number estimates for (stable) integral simplicial volume and integral
foliated simplicial volume can be obtained through Poincaré duality:

Proposition 6.4.3 (integral Betti number estimate [75]). *Let M be an oriented
closed connected manifold and let $k \in \mathbb{N}$. Then*

$$b_k(M) \leq \|M\|_{\mathbb{Z}}.$$

Proof. Let $c = \sum_{j=1}^{m} a_j \cdot \sigma_j \in Z(M; \mathbb{Z})$ (in reduced form). Then, by Poincaré
duality, the cap product

$$\cdot \cap [M]_{\mathbb{Z}} \colon H^{n-k}(M; \mathbb{Q}) \longrightarrow H_k(M; \mathbb{Q})$$

$$[f] \longmapsto (-1)^{(n-k) \cdot k} \cdot \left[\sum_{j=1}^{m} a_j \cdot f(_{n-k}\lfloor \sigma_j) \cdot \sigma_j \rfloor_k \right]$$

is an isomorphism of $\mathbb{Q}$-vector spaces. In particular, $H_k(M; \mathbb{Q})$ is a quotient
of a subspace of a $\mathbb{Q}$-vector space that is generated by m elements (namely,
$\sigma_1 \rfloor_k, \ldots, \sigma_m \rfloor_k$). Because the coefficients of c are integral, we obtain

$$b_k(M) = \dim_{\mathbb{Q}} H_k(M; \mathbb{Q}) \leq m \leq |c|_1.$$

Taking the infimum over all c in $Z(M; \mathbb{Z})$, we obtain $b_k(M) \leq \|M\|_{\mathbb{Z}}$.   □

Corollary 6.4.4 (L^2-Betti number estimate, residually finite case). *Let M be an
oriented closed connected manifold with residually finite fundamental group
and let $k \in \mathbb{N}$. Then*

$$b_k^{(2)}(M) \leq \|M\|_{\mathbb{Z}}^{\infty}.$$

Proof. If Γ_* is a residual chain of $\Gamma := \pi_1(M)$, then (check!)

$$\inf_{n \in \mathbb{N}} \frac{\|M_n\|_{\mathbb{Z}}}{[\Gamma : \Gamma_n]} = \lim_{n \to \infty} \frac{\|M_n\|_{\mathbb{Z}}}{[\Gamma : \Gamma_n]},$$

where M_n denotes the finite cover associated with the subgroup $\Gamma_n \subset \Gamma$. Hence, the previous Betti number estimate (Proposition 6.4.3) and the approximation theorem (Theorem 3.1.3) show that

$$b_k^{(2)}(M) \le \inf_{n \in \mathbb{N}} \frac{\|M_n\|_{\mathbb{Z}}}{[\Gamma : \Gamma_n]}.$$

Taking the infimum over all residual chains of Γ finishes the proof.   □

Theorem 6.4.5 (L^2-Betti number estimate, general case [126, Corollary 5.28]). *Let M be an oriented closed connected manifold and let $k \in \mathbb{N}$. Then*

$$b_k^{(2)}(M) \le |M|.$$

Proof. We will proceed as in the proof of Proposition 6.4.3, replacing the ring $\mathbb{Z}$ with its ergodic theoretic analogue. Let $\Gamma \curvearrowright X$ be an essentially free standard Γ-space and let $\mathcal{R} := \mathcal{R}_{\Gamma \curvearrowright X}$ be the associated orbit relation; such an action exists, as discussed in the proof of Theorem 4.3.10. The trace-preserving $*$-homomorphism $N\Gamma \hookrightarrow N\mathcal{R}$ shows that $N\mathcal{R}$ is a flat $N\Gamma$-module (Remark 4.2.7) and hence that

$$
\begin{aligned}
b_k^{(2)}(M) &= \dim_{N\Gamma} H_k(M; N\Gamma) && \text{([105, Lemma 6.53])} \\
&= \dim_{N\mathcal{R}} N\mathcal{R} \otimes_{N\Gamma} H_k(M; N\Gamma) && \text{(Remark 4.2.7)} \\
&= \dim_{N\mathcal{R}} H_k(M; N\mathcal{R}). && \text{($N\mathcal{R}$ is flat over $N\Gamma$)}
\end{aligned}
$$

We now make use of twisted Poincaré duality: Let $c \in Z(M; L^\infty(X; \mathbb{Z}))$, say $c = \sum_{j=1}^m f_j \otimes \sigma_j$ (in reduced form). Then

$$\cdot \cap [c] \colon H^{n-k}(M; N\mathcal{R}) \longrightarrow H_k(M; N\mathcal{R})$$

$$[f] \longmapsto (-1)^{(n-k)\cdot k} \cdot \left[\sum_{j=1}^m \overline{f(_{n-k}\lfloor \sigma_j)} \cdot \overline{f_j} \otimes \sigma_j \rfloor_k \right]$$

is an isomorphism of $N\mathcal{R}$-modules [126, Corollary 5.17], where $\overline{}$ denotes the involution on $N\mathcal{R}$. Then $H_k(M; N\mathcal{R})$ is isomorphic to a quotient of a submodule of the $N\mathcal{R}$-module $\bigoplus_{j=1}^m N\mathcal{R} \cdot \chi_{\mathrm{supp}\,\overline{f_j}} = \bigoplus_{j=1}^m N\mathcal{R} \cdot \overline{\chi_{\mathrm{supp}\,f_j}}$. Therefore, the additivity properties of $\dim_{N\mathcal{R}}$ and Lemma 4.3.12 show that

$$
\begin{aligned}
b_k^{(2)}(M) &= \dim_{N\mathcal{R}} H_k(M; N\mathcal{R}) \\
&\le \dim_{N\mathcal{R}} \left(\bigoplus_{j=1}^m N\mathcal{R} \cdot \chi_{\mathrm{supp}\,\overline{f_j}} \right) = \sum_{j=1}^m \mu(\mathrm{supp}\, f_j) \\
&\le |c|_1;
\end{aligned}
$$

again, it is essential that the coefficient functions are $\mathbb{Z}$-valued. Taking the infimum over all c in $Z(M; L^\infty(X, \mathbb{Z}))$, we obtain the desired estimate.   □

Similar estimates for L^2-Betti numbers via parametrised norms have been combined with randomised covers and nerve retracts to establish upper bounds for L^2-Betti numbers in terms of the volume and in the case of amenable covers with controlled multiplicity [124].

6.4.3 The rank gradient/cost estimate

In degree 1, the $(L^2$-)Betti number estimates admit the following non-commutative refinements:

Theorem 6.4.6 (rank gradient estimate [98]). *Let M be an oriented closed connected manifold with residually finite (infinite) fundamental group. Then*

$$\mathrm{rg}\big(\pi_1(M)\big) \leq \|M\|_{\mathbb{Z}}^{\infty}.$$

Proof. In view of stabilisation along finite index subgroups/finite coverings, it suffices to prove the following estimate:

$$d\big(\pi_1(M)\big) \leq \|M\|_{\mathbb{Z}}.$$

Let $c \in Z(M;\mathbb{Z})$, let $\Gamma := \pi_1(M)$, and let $n := \dim M$. We then take the lift $\tilde{c} = \sum_{j=1}^{m} a_j \cdot \tilde{\sigma}_j \in C_n^{\mathrm{sing}}(\widetilde{M};\mathbb{Z})$ (in reduced form) of c to the universal covering $\widetilde{M}$ that satisfies $a_1, \ldots, a_m \in \mathbb{Z}$ and $\tilde{\sigma}_1(v_0), \ldots, \tilde{\sigma}_m(v_0) \in D$, where D is a chosen strict fundamental domain of the deck transformation action $\Gamma \curvearrowright \widetilde{M}$ and v_0 is the 0-vertex of Δ^n. For $j \in \{1, \ldots, m\}$, let $\gamma_j \in \Gamma$ be the unique group element with $\gamma_j \cdot \sigma_j(v_1) \in D$; we then consider the subgroup

$$\Lambda := \langle \gamma_1, \ldots, \gamma_m \rangle_{\Gamma} \subset \Gamma.$$

By construction, $d(\Lambda) \leq m \leq |c|_1$ and it suffices to show that $\Lambda = \Gamma$.

Let $\pi_\Lambda \colon \overline{M} := \Lambda \setminus \widetilde{M} \longrightarrow M$ be the covering of M associated with Λ and let $p_\Lambda \colon \widetilde{M} \longrightarrow \Lambda \setminus \widetilde{M} = \overline{M}$ be the corresponding "upper" covering map. Then the chain

$$\overline{c} := C_n^{\mathrm{sing}}(p_\Lambda;\mathbb{Z})(\tilde{c}) \in C_n^{\mathrm{sing}}(\overline{M};\mathbb{Z})$$

is a cycle in $C_*^{\mathrm{sing}}(\overline{M};\mathbb{Z})$ (check!) and

$$H_n(\pi_\Lambda;\mathbb{Z})\big([\overline{c}]\big) = [c] = [M]_{\mathbb{Z}} \in H_n(M;\mathbb{Z}).$$

Therefore, $H_n(\overline{M};\mathbb{Z}) \neq 0$ and we obtain $[\Gamma : \Lambda] = |\deg \pi_\Lambda| = 1$.

Alternatively, there is also a geometric argument via combinatorial models of cycles and covering theory that proves the slightly weaker bound $\mathrm{rg}\,\Gamma \leq \dim M \cdot \|M\|_{\mathbb{Z}}^{\infty}$ [98, Lemma 4.1]. $\square$

By Theorem 4.3.5 and Theorem 6.4.1, this rank gradient estimate could also be derived from the dynamical version of Theorem 6.4.6:

Theorem 6.4.7 (cost esimate [99]). *Let M be an oriented closed connected manifold with fundamental group Γ and let $\Gamma \curvearrowright X$ be an essentially free ergodic standard Γ-action. Then*

$$\mathrm{cost}(\Gamma \curvearrowright X) - 1 \leq |M|^{\Gamma \curvearrowright X}.$$

In particular, $\mathrm{cost}\,\Gamma - 1 \leq |M|$ (Remark 4.3.3).

Proof. If Γ is finite, then the left-hand side is non-positive (Exercise 4.E.8). Therefore, it suffices to consider the case when Γ is infinite.

Let $c \in Z(M; L^\infty(X, \mathbb{Z}))$; decomposing the coefficient functions of c into their finitely many steps, we may write c in the (reduced) form $c = \sum_{j=1}^m a_j \cdot \chi_{A_j} \otimes \sigma_j$, where $a_1, \ldots, a_m \in \mathbb{Z} \setminus \{0\}$, $A_1, \ldots, A_m \subset X$ are measurable subsets and $\sigma_1, \ldots, \sigma_m$ are singular simplices on $\widetilde{M}$ with the property that $\sigma_1(v_0), \ldots, \sigma_m(v_0)$ lie in the same (strict) fundamental domain D of $\Gamma \curvearrowright \widetilde{M}$. We then consider for $j \in \{1, \ldots, m\}$ the unique element $\gamma_j \in \Gamma$ with $\gamma_j \cdot \sigma_j(v_1) \in D$ and the associated partial translation automorphism

$$\varphi_j := \gamma_j \cdot : A_j \longrightarrow \gamma_j \cdot A_j$$

of $\mathcal{R} := \mathcal{R}_{\Gamma \curvearrowright X}$. Let $R := \langle \varphi_1, \ldots, \varphi_m \rangle \subset \mathcal{R}$ be the subrelation generated by these partial automorphisms; this is a dynamical version of the subgroup considered in the proof of Theorem 6.4.6.

In general, R will not necessarily be a finite index subrelation of $\mathcal{R}$ [99, Remark 4.2], but the inclusion $R \subset \mathcal{R}$ will always be a so-called translation finite extension [99, Definition 2.9, Lemma 4.5]. Therefore, we obtain

$$\mathrm{cost}\,\mathcal{R} - 1 \leq \mathrm{cost}\,R \qquad \text{(by translation finiteness [99, Lemma 2.11])}$$

$$\leq \sum_{j=1}^m \mu(A_j) \qquad ((\varphi_j)_{j \in \{1, \ldots, m\}} \text{ is a graphing of } R)$$

$$\leq |c|_1.$$

We can now take the infimum over all $c \in Z(M; L^\infty(X, \mathbb{Z}))$. $\qquad\qquad \square$

6.4.4 Amenable fundamental group

Theorem 6.4.8 (amenable fundamental group, residually finite case [60, Theorem 1.10]). *Let M be an oriented closed connected aspherical manifold of nonzero dimension whose fundamental group is residually finite and amenable. Then*

$$\|M\|_{\mathbb{Z}}^\infty = 0.$$

The classical proof of vanishing of ordinary simplicial volume of manifolds with amenable fundamental group relies on invariant means and bounded cohomology. These *co*homological methods are not available in the residually finite or the dynamical view, but for *aspherical* manifolds we can replace the cohomological averaging by a Følner argument on the chain level.

Remark 6.4.9 (amenability via Følner sequences). Let Γ be a finitely generated group with finite generating set S. If Γ is residually finite and amenable, then there exists a *Følner sequence* (with respect to S), i.e., a sequence $(F_k)_{k\in\mathbb{N}}$ of non-empty finite subsets of Γ with

$$\lim_{k\to\infty} \frac{|\partial_S F_k|}{|F_k|} = 0.$$

Here, $\partial_S F := \{\gamma \in F \mid \exists_{s\in S} \quad \gamma \cdot s \notin F\}$ denotes the *S-boundary* of a subset $F \subset \Gamma$.

Moreover, this sequence can be chosen in such a way that there is a residual chain $(\Gamma_k)_{k\in\mathbb{N}}$ of Γ with the property that, for each $k \in \mathbb{N}$, the set F_k is a set of coset representatives for Γ_k in Γ [45, Proposition 5.5].

Proof of Theorem 6.4.8. Let Γ be the fundamental group of M, let $n :=$ $\dim M$, and let $c \in Z(M;\mathbb{Z})$. We proceed as follows:

- We lift c to a chain $\tilde{c} \in C_n^{\mathrm{sing}}(\widetilde{M};\mathbb{Z})$. This chain will not be a cycle; the defect of $\tilde{c}$ of being a cycle is measured by a subset $S \subset \Gamma$.

- We choose a Følner sequence $(F_k)_{k\in\mathbb{N}}$ of Γ for S as in Remark 6.4.9.

- For each $k \in \mathbb{N}$, we bound $|\partial(F_k \cdot \tilde{c})|_1$ linearly in terms of $|\partial_S F_k|$.

- We then efficiently fill $\partial(F_k \cdot \tilde{c})$ by a new chain $\tilde{z}_k$ (using that $\widetilde{M}$ is contractible).

- Pushing down $\tilde{z}_k$ to the finite covering manifold $M_k := \Gamma_k \setminus \widetilde{M}$ of M leads to a cycle $z_k \in Z(M_k;\mathbb{Z})$ with $|z_k|_1 \leq \mathrm{const}\cdot|\partial_S F_k|/|F_k| \cdot |c|_1$.

- Taking $k \to \infty$ then shows that $\|M\|_{\mathbb{Z}}^\infty = 0$.

We now add details: We write $c = \sum_{j=1}^m a_j \cdot \sigma_j$ with $a_1,\ldots,a_m \in \{-1,1\}$ and $m = |c|_1$. Let $\tilde{c} = \sum_{j=1}^m a_j \cdot \tilde{\sigma}_j \in C_n^{\mathrm{sing}}(\widetilde{M};\mathbb{Z})$ be a lift of c. As c is a cycle, there is a matching of the set of faces of $\sigma_1,\ldots,\sigma_m$ that appear with positive sign in the expression ∂c and the set of faces of $\sigma_1,\ldots,\sigma_m$ that appear with negative sign. Let $\Sigma = \Sigma_+ \sqcup \Sigma_-$ be a corresponding splitting of the set of all faces of $\sigma_1,\ldots,\sigma_m$. For $\tau \in \Sigma$, we denote the matched face by τ^- and the corresponding face in the lift $\tilde{c}$ by $\tilde{\tau}$. By the cancellation, for each $\tau \in \Sigma_+$, there exists a $\gamma_\tau \in \Gamma$ with $\tilde{\tau}^- = \gamma_\tau \cdot \tilde{\tau}$. Hence,

$$\partial\tilde{c} = \sum_{\tau\in\Sigma_+} (\tilde{\tau} - \gamma_\tau \cdot \tilde{\tau}).$$

Let $S \subset \Gamma$ be a symmetric finite generating set that contains $\{\gamma_\tau \mid \tau \in \Sigma_+\}$.

Because Γ is amenable, there exists a Følner sequence $(F_k)_{k\in\mathbb{N}}$ of Γ with respect to S as in Remark 6.4.9. For $k \in \mathbb{N}$, let $F_k \cdot \widetilde{c} := \sum_{\gamma\in F_k} \gamma \cdot \widetilde{c}$. By construction, for each $k \in \mathbb{N}$, we have

$$\left|\partial(F_k \cdot \widetilde{c})\right|_1 \leq \sum_{\tau\in\Sigma_+}\left|\sum_{\gamma\in F_k}\gamma\cdot\widetilde{\tau} - \sum_{\gamma\in F_k}\gamma\cdot\gamma_k\cdot\widetilde{\tau}\right| \leq |\Sigma_+|\cdot 2\cdot|\partial_S F_k|$$
$$\leq |c|_1 \cdot |\partial_S F_k|.$$

Because $\widetilde{M}$ is contractible, a cone construction shows that there exists a chain $\widetilde{z}_k \in C_n^{\mathrm{sing}}(\widetilde{M};\mathbb{Z})$ that satisfies [54, Lemma 4.1][60, Lemma 6.3]

$$\partial\widetilde{z}_k = \partial(F_k\cdot\widetilde{c}) \quad\text{and}\quad |\widetilde{z}_k|_1 \leq \left|\partial(F_k\cdot\widetilde{c})\right|_1 \leq |c|_1\cdot|\partial_S F_k|.$$

We write $q_k\colon \widetilde{M} \longrightarrow M_k = \Gamma_k\backslash\widetilde{M}$ for the "upper" covering associated with Γ_k. Then

$$z_k := C_n^{\mathrm{sing}}(q_k;\mathbb{Z})(\widetilde{z}_k) \in C_n^{\mathrm{sing}}(M_k;\mathbb{Z})$$

is a cycle: The chain $\widetilde{z}_k - F_k\cdot\widetilde{c}$ is an n-cycle on the contractible space $\widetilde{M}$, whence null-homologous; moreover, F_k is a set of coset representatives for Γ_k in Γ. The cycle z_k is even a fundamental cycle of M_k (because it pushes down to $[\Gamma:\Gamma_k]$-times c on M). We then conclude that

$$\|M\|_{\mathbb{Z}}^{\infty} \leq \frac{1}{[\Gamma:\Gamma_k]}\cdot\|M_k\|_{\mathbb{Z}} \leq \frac{1}{[\Gamma:\Gamma_k]}\cdot|z_k|_1 \leq \frac{1}{|F_k|}\cdot|\partial_S F_k|\cdot|c|_1.$$

As $(F_k)_{k\in\mathbb{N}}$ is a Følner sequence, the right-hand side tends to 0. $\square$

Theorem 6.4.10 (amenable fundamental group, general case [60, Theorem 1.9]). *Let M be an oriented closed connected aspherical manifold of non-zero dimension with amenable fundamental group. Then M is cheap, i.e., for every essentially free standard $\pi_1(M)$-space α, we have*

$$|M|^{\alpha} = 0.$$

In particular, $|M| = 0$.

Proof. The proof is a dynamical version of the proof of Theorem 6.4.8; we use the same notation as in that proof and modify the construction of the improved cycles as follows. Instead of passage to finite coverings and division by the covering degree in the stabilisation step, we perform division inside the probability space via the following version of the Rokhlin lemma by Ornstein and Weiss [124, Theorem 5.2]:

Let $\alpha = (\Gamma \curvearrowright (X,\mu))$ and let $\varepsilon \in \mathbb{R}_{>0}$. Then there exists an $N \in \mathbb{N}$, finite subsets $F_1,\dots,F_N \subset \Gamma$, and measurable sets $A_1,\dots,A_N \subset X$ such that

- for each $j \in \{1,\dots,N\}$, the set F_j satisfies $|\partial_S F_j|/|F_j| \leq \varepsilon$;

- for each $j \in \{1, \dots, N\}$, the sets $\gamma \cdot A_j$ with $\gamma \in F_j$ are pairwise disjoint and the sets $F_j \cdot A_j$ with $j \in \{1, \dots, N\}$ are pairwise disjoint;

- the rest $R := X \setminus \bigcup_{j=1}^{N} F_j \cdot A_j$ satisfies $\mu(R) < \varepsilon$.

For $j \in \{1, \dots, N\}$, we set $\widetilde{c}_j := F_j^{-1} \cdot \widetilde{c} \in C_n^{\mathrm{sing}}(\widetilde{M}; \mathbb{Z})$. We then choose a filling $\widetilde{z}_j \in C_n^{\mathrm{sing}}(\widetilde{M}; \mathbb{Z})$ with $\partial \widetilde{z}_j = \partial \widetilde{c}_j$ and $|\widetilde{z}_j|_1 \leq |c|_1 \cdot |\partial_S F_j|$. Finally, we set

$$z := \sum_{j=1}^{N} \chi_{A_j} \otimes \widetilde{z}_j + \sum_{j=1}^{m} a_j \cdot \chi_R \otimes \widetilde{\sigma}_j \in C_n^{\mathrm{sing}}(M; L^\infty(X, \mathbb{Z})).$$

Then a straightforward calculation shows that $z \in Z(M; L^\infty(X, \mathbb{Z}))$ [60, Lemma 6.4] and that

$$|z|_1 \leq \sum_{j=1}^{N} \mu(A_j) \cdot |\widetilde{z}_j|_1 + \mu(R) \cdot m \leq \sum_{j=1}^{N} \mu(A_j) \cdot |c|_1 \cdot \varepsilon \cdot |F_j| + \mu(R) \cdot |c|_1$$

$$\leq 1 \cdot \varepsilon \cdot |c|_1 + \varepsilon \cdot |c|_1.$$

Taking $\varepsilon \to 0$ proves the claim. □

In combination with the (L^2)-Betti number estimates (Chapter 6.4.2) and rank gradient estimates (Chapter 6.4.3), the vanishing results Theorem 6.4.8 and Theorem 6.4.10 give alternative and unified proofs for vanishing results for L^2-Betti numbers, Betti number gradients, rank gradients, cost, and log-torsion homology gradients of closed aspherical manifolds with (residually finite) amenable fundamental group [60].

6.4.5   Hyperbolic 3-manifolds

Theorem 6.4.11 (hyperbolic 3-manifolds [60, Theorem 1.7]). *Let M be an oriented closed connected hyperbolic 3-manifold. Then*

$$\|M\|_{\mathbb{Z}}^\infty = \|M\| = \frac{\mathrm{vol}(M)}{v_3},$$

where v_3 is the volume of a (whence every) ideal regular tetrahedron in $\mathbb{H}^3$.

The proof uses concrete geometric examples, a version of Thurston's smearing construction, and ergodic theory:

Remark 6.4.12 (concrete examples). Let $(M_n)_{n \in \mathbb{N}}$ be a sequence of (distinct) oriented closed connected hyperbolic 3-manifolds obtained by Dehn surgery on the 5-chain link complement $M_{(5)}$. Then Thurston's Dehn filling theorem, a concrete ideal triangulation of $M_{(5)}$, and estimates in terms of the complexity show that [100, Theorem 1.6]

$$\lim_{n\to\infty} \frac{\|M_n\|_{\mathbb{Z}}^{\infty}}{\|M_n\|} = 1.$$

Proposition 6.4.13 (a weak proportionality principle [55, Theorem 4.1]). *Let M be an oriented closed connected hyperbolic n-manifold. Then, for every oriented closed connected hyperbolic n-manifold N, we obtain*

$$\frac{|M|}{\text{vol}(M)} \le \frac{\|N\|_{\mathbb{Z}}^{\infty}}{\text{vol}(N)}.$$

Sketch of proof. By definition of $\|\cdot\|_{\mathbb{Z}}^{\infty}$ and the multiplicativity of Riemannian volume with respect to finite coverings, it suffices to show that

$$\frac{|M|}{\text{vol}(M)} \le \frac{\|N\|_{\mathbb{Z}}}{\text{vol}(N)}.$$

We use a dynamical version of the discrete smearing map: Let $\pi_M \colon \mathbb{H}^n \longrightarrow M$ be the universal covering map of M.

Let Γ and Λ be the fundamental groups of M and N, respectively. These groups are lattices in $G := \text{Isom}^+(\mathbb{H}^n)$. We consider the standard Γ-space α given by the left translation action on G/Λ and the normalised Haar measure. We choose a sufficiently fine Γ-equivariant Borel partition of $\mathbb{H}^n$ and a Γ-net of points in these sets. Let $S_k \subset \text{map}(\Delta^k, M)$ be the set of singular simplices that lift to geodesic simplices on $\mathbb{H}^n$ with vertices in this net. Moreover, we choose a π_M-lift $\widetilde{\varrho}$ for each $\varrho \in S_k$.

For a singular k-simplex $\overline{\tau}$ on $\mathbb{H}^n$, we write $\text{snap}(\tau)$ for the geodesic k-simplex in $\Gamma \cdot \widetilde{S}_k$ on $\mathbb{H}^n$ that is obtained by "snapping" the vertices of τ to the points in the Γ-net that lie in the same Borel set and subsequent geodesic straightening of the simplex. We then consider the map (where $\widetilde{\sigma}$ is any choice of a lift of σ to $\mathbb{H}^n$)

$$\varphi_k \colon C_k^{\text{sing}}(N;\mathbb{Z}) \longrightarrow C_k^{\text{sing}}\big(M; L^{\infty}(G/\Lambda, \mathbb{Z})\big)$$
$$\text{map}(\Delta^k, N) \ni \sigma \longmapsto \sum_{\varrho \in S_k} \big(x \cdot \Lambda \mapsto \big|\{\lambda \in \Lambda \mid \text{snap}(x \cdot \lambda \cdot \widetilde{\sigma} = \widetilde{\varrho}\}\big|\big) \otimes \varrho$$

that "smears" simplices over $\mathbb{H}^n$ via G. Integration and comparison with the classical smearing map shows [55, Section 4]: For all $c \in Z(N;\mathbb{Z})$, we have $\varphi_n(c) \in Z(M; L^{\infty}(G/\Lambda, \mathbb{Z}))$ and

$$|\varphi_n(c)|_1 \le \frac{\text{vol}(M)}{\text{vol}(N)} \cdot |c|_1.$$

Therefore,

$$\frac{|M|}{\text{vol}(M)} \le \frac{|M|^{\alpha_M}}{\text{vol}(M)} \le \frac{\|N\|_{\mathbb{Z}}}{\text{vol}(N)}. \qquad \square$$

Remark 6.4.14 (monotonicity with respect to weak containment). Let Γ be a countable group and let $\alpha = \Gamma \curvearrowright (X, \mu)$ and $\beta = \Gamma \curvearrowright (Y, \nu)$ be standard Γ-spaces. Then α is *weakly contained* in β, in symbols $\alpha \prec \beta$, if: For every $\varepsilon \in \mathbb{R}_{>0}$, every finite subset $F \subset \Gamma$, every $m \in \mathbb{N}$, and all measurable sets $A_1, \ldots, A_m \subset X$, there exist measurable sets $B_1, \ldots, B_m \subset Y$ with

$$\forall_{\gamma \in F} \ \forall_{j,k \in \{1,\ldots,m\}} \ \ \left| \mu(\gamma^\alpha(A_j) \cap A_k) - \mu(\gamma^\beta(B_j) \cap B_k) \right| < \varepsilon.$$

If Γ is infinite, then every essentially free standard Γ-space weakly contains the Bernoulli shift of Γ [8]. Moreover, an infinite finitely generated residually finite group Γ is said to have property EMD*, if every standard Γ-space is weakly contained in the profinite completion $\Gamma \curvearrowright \widehat{\Gamma}$ of Γ [88].

If M is an oriented closed connected manifold with fundamental group Γ and if α and β are essentially free non-atomic standard Γ-spaces with $\alpha \prec \beta$, then [60, Theorem 3.3]

$$|M|^{\,\beta} \leq |M|^{\,\alpha}.$$

In particular: If Γ satisfies EMD*, then $\|M\|_{\mathbb{Z}}^\infty = |M|$.

Proof of Theorem 6.4.11. We already know that $\|M\|_{\mathbb{Z}}^\infty \geq \|M\|$ and $\|M\| = \mathrm{vol}(M)/v_3$ (by the computation of Gromov and Thurston); hence, it remains to show that $\|M\|_{\mathbb{Z}}^\infty \leq \|M\|$.

As observed by Kechris, Bowen, and Tucker-Drob, in the hyperbolic case, $\pi_1(M)$ satisfies EMD* [60, Proposition 3.10]. Therefore, it suffices to show that $|M| \leq \|M\|$ (Remark 6.4.14). Let $(M_n)_{n \in \mathbb{N}}$ be a sequence of oriented closed connected 3-manifolds as in Remark 6.4.12. In combination with the weak proportionality principle (Proposition 6.4.13) and the proportionality principle for ordinary simplicial volume, we obtain

$$|M| \leq \lim_{n \to \infty} \mathrm{vol}(M) \cdot \frac{\|M_n\|_{\mathbb{Z}}^\infty}{\mathrm{vol}(M_n)} = \lim_{n \to \infty} \|M\| \cdot \frac{\|M_n\|_{\mathbb{Z}}^\infty}{\|M_n\|} = \|M\|,$$

as desired. $\qquad\qquad\qquad\qquad\qquad\qquad\qquad\qquad\qquad\qquad\qquad\qquad\qquad$ $\square$

Outlook 6.4.15 (profinite rigidity). A prominent open problem in 3-manifold topology is to decide whether two oriented closed connected hyperbolic 3-manifolds M and N with fundamental group Γ and Λ, respectively, satisfy $\widehat{\Gamma} \cong \widehat{\Lambda}$ if and only if $\mathrm{vol}(M) = \mathrm{vol}(N)$.

By Theorem 6.4.1 and Theorem 6.4.11, we have $|M|^{\,\Gamma \curvearrowright \widehat{\Gamma}} = \mathrm{vol}(M)/v_3$. However, it is not clear whether $|M|^{\,\Gamma \curvearrowright \widehat{\Gamma}}$ can be recovered from the profinite completion $\widehat{\Gamma}$ alone. In this context, it should be noted that the first L^2-Betti number is a profinite invariant for finitely presented residually finite groups [29], but that the higher L^2-Betti numbers, in general, are *not* profinite invariants [85].

The proportionality principle in Proposition 6.4.13 can be refined as follows [100, Corollary 1.3]: Let X be an aspherical symmetric space of non-compact type and let $\Gamma, \Lambda \subset \mathrm{Isom}^0(X)$ be uniform torsion-free lattices. Then

$$\frac{|\Gamma \backslash X|}{\mathrm{covol}(\Gamma)} = \frac{|\Lambda \backslash X|}{\mathrm{covol}(\Lambda)}.$$

The proof uses the bounded mixing measure equivalence given by $\mathrm{Isom}^0(X)$ between Γ and Λ and a suitable version of the associated ME-cocycle, similar to work of Bader, Furman, and Sauer [14] (Exercise 4.E.10).

6.4.6   Aspherical 3-manifolds

Theorem 6.4.16 (aspherical 3-manifolds [55, Theorem 1]). *Let M be an oriented closed connected aspherical 3-manifold. Then (where* hypvol *is the sum of volumes of hyperbolic pieces in the JSJ-decomposition of M)*

$$\|M\|_{\mathbb{Z}}^\infty = \|M\| = \frac{\mathrm{hypvol}(M)}{v_3}.$$

Sketch of proof. Work of Soma (and geometrisation) shows that $\|M\| = \mathrm{hypvol}(M)/v_3$ [128]. Moreover, we always have $\|M\| \leq \|M\|_{\mathbb{Z}}^\infty$; therefore, it suffices to show that $\|M\|_{\mathbb{Z}}^\infty \leq \mathrm{hypvol}(M)/v_3$. We first decompose M via geometrisation. Because M is closed and aspherical, M admits a JSJ-decomposition, i.e., we can dissect M along π_1-injective tori into Seifert fibred pieces and hyperbolic pieces.

The proof of Theorem 6.4.11 generalises to the case with toroidal boundary [55, Corollary 5.3]: If $(W, \partial W)$ is an oriented compact connected 3-manifold with empty or toroidal boundary, whose interior admits a hyperbolic structure, then

$$|W, \partial W|_\partial^{\widehat{\pi_1}(W)} = \frac{\mathrm{vol}(W^\circ)}{v_3}.$$

Here, $|W, \partial W|_\partial$ is a relative version of integral foliated simplicial volume that includes control on the norm of the boundaries of relative fundamental cycles. This boundary control, together with a Følner type argument, allows us to establish sub-additivity along JSJ-tori [55, Proposition 6.4/6.5] (this phenomenon is similar to Example 4.3.9). Therefore, we obtain:

$$\|M\|_{\mathbb{Z}}^\infty = |M|^{\widehat{\pi_1}(M)} \qquad\qquad\qquad \text{(Theorem 6.4.1)}$$

$$\leq \sum_{W \text{ JSJ-piece of } M} |W, \partial W|_\partial^{\widehat{\pi_1}(M)} \quad \text{(additivity along JSJ-tori)}$$

If $(W, \partial W)$ is a JSJ-piece of M, then it is known from 3-manifold theory that $\widehat{\pi_1}(W)$ is weakly contained in $\widehat{\pi_1}(M)$ [55, Section 6.3]. Hence, the relative version of Remark 6.4.14 [55, Proposition A.1] shows that

$$|W, \partial W|_{\partial}^{\widehat{\pi_1}(M)} \leq |W, \partial W|_{\partial}^{\widehat{\pi_1}(W)}.$$

If $(W, \partial W)$ is Seifert fibred, then $|W, \partial W|_{\partial}^{\widehat{\pi_1}(W)} = \|W, \partial W\|_{\mathbb{Z}}^{\infty} = 0$ [100, Section 8]. Hence, it follows that

$$\|M\|_{\mathbb{Z}}^{\infty} \leq \sum_{W \text{ hyp. JSJ-piece of } M} |W, \partial W|_{\partial}^{\widehat{\pi_1}(W)}$$

$$= \sum_{W \text{ hyp. JSJ-piece of } M} \frac{\text{vol}(W^{\circ})}{v_3} = \frac{\text{hypvol}(M)}{v_3}. \qquad \square$$

Remark 6.4.17 (non-approximation for non-aspherical 3-manifolds [55, Section 1.3]). General closed 3-manifolds (with infinite fundamental group) do *not* satisfy approximation for simplicial volume. For example, the first L^2-Betti number can be used to detect this:

Let N be an oriented closed connected hyperbolic 3-manifold and let $k \in \mathbb{N}$ with $k > \text{vol}(N)/v_3$. Then the oriented closed connected 3-manifold $M := N \# \#^k (S^1)^3$ satisfies $\|M\| < \|M\|_{\mathbb{Z}}^{\infty}$: On the one hand, the additivity of simplicial volume with respect to connected sums in dimension bigger than 2 [73, p. 10] and the computation of simplicial volume of hyperbolic 3-manifolds show that

$$\|M\| = \|N\| + k \cdot \|(S^1)^3\| = \frac{\text{vol}(N)}{v_3} < k.$$

On the other hand, we obtain from the additivity properties of the first L^2-Betti number [105, Theorem 1.35] and from the L^2-Betti number estimate (Corollary 6.4.4) that

$$\|M\|_{\mathbb{Z}}^{\infty} \geq b_1^{(2)}(M) = b_1^{(2)}(N) + k = k > \|M\|.$$

6.E Exercises

Exercise 6.E.1 (multiplicativity of simplicial volumes). Let M be an oriented closed connected manifold and let $N \longrightarrow M$ be a d-sheeted (finite) covering of M.

1. Show that $\|M\| \leq 1/d \cdot \|N\|$ (using push-forwards of fundamental cycles of N).

2. Show that $\|N\| \leq d \cdot \|M\|$ and $\|N\|_{\mathbb{Z}} \leq d \cdot \|M\|_{\mathbb{Z}}$ (using the transfer of fundamental cycles of M).

3. Conclude that $\|N\|_{\mathbb{Z}}^{\infty} = d \cdot \|M\|_{\mathbb{Z}}^{\infty}$.

Exercise 6.E.2 (simplicial volume of spheres and tori). Use self-maps of non-trivial degree to show that spheres and tori in non-zero dimension have simplicial volume equal to 0. What about stable integral simplicial volume?

Exercise 6.E.3 (simplicial volumes of surfaces [73]). Let $g \in \mathbb{N}_{\geq 2}$ and let Σ_g be "the" oriented closed connected surface of genus g.

1. Use explicit (singular) triangulations of surfaces and covering theory to prove that

$$\|\Sigma_g\| \leq \|\Sigma_g\|_{\mathbb{Z}}^{\infty} \leq 4 \cdot (g-1) = 2 \cdot |\chi(\Sigma_g)|.$$

2. Use integration of (smooth, geodesically straightened) singular simplices and hyperbolic geometry to prove that

$$2 \cdot |\chi(\Sigma_g)| \leq \|\Sigma_g\|.$$

Exercise 6.E.4 (vanishing stable integral simplicial volume?!).

1. Does there exist an oriented closed connected manifold M satisfying $\pi_1(M) \cong F_2$ and $\|M\|_{\mathbb{Z}}^{\infty} = 0$?!

2. Does there exist an oriented closed connected manifold M satisfying $\pi_1(M) \cong F_2 \times F_2$ and $\|M\|_{\mathbb{Z}}^{\infty} = 0$?!

Hints. (L^2-)Betti numbers might help ...

Exercise 6.E.5 (simplicial volume, L^2-Betti numbers, and the Singer conjecture). Let M be an oriented closed connected aspherical manifold. How are Question 6.1.2 and the Singer conjecture (Outlook 2.2.9) related?

A
Quick reference

We collect basic terminology on some objects used in the main text.

Overview of this chapter.

A.1   Von Neumann algebras   94
A.2   Weak convergence of measures   94
A.3   Lattices   95

© The Author(s), under exclusive license to Springer Nature Switzerland AG 2020   93
C. Löh, *Ergodic Theoretic Methods in Group Homology*,
SpringerBriefs in Mathematics, https://doi.org/10.1007/978-3-030-44220-0

A.1 Von Neumann algebras

If H is a complex Hilbert space, then $B(H)$ denotes the space of all bounded linear operators $H \longrightarrow H$. Then $B(H)$ is a $\mathbb{C}$-algebra with respect to the pointwise linear structure and the multiplication given by composition. Moreover, $B(H)$ carries an involution, given by taking adjoints. This turns $B(H)$ into a $*$-algebra. There are several topologies on $B(H)$:

- A net $(T_i)_{i \in I}$ in $B(H)$ *converges in operator norm* to $T \in B(H)$ if the net $(\|T - T_i\|)_{i \in I}$ of operator norms converges to 0.

- A net $(T_i)_{i \in I}$ in $B(H)$ *converges strongly* to $T \in B(H)$ if for each $x \in H$, the net $(T_i(x))_{i \in I}$ converges in H (in norm).

- A net $(T_i)_{i \in I}$ in $B(H)$ *converges weakly* to $T \in B(H)$ if for each $x \in H$, the net $(T_i(x))_{i \in I}$ converges weakly in H.

Definition A.1.1 (von Neumann algebra). A *von Neumann algebra* is a unital weakly closed $*$-subalgebra of $B(H)$ for some complex Hilbert space H.

Theorem A.1.2 (von Neumann bicommutant theorem [57, Corollary 4.2.2]). *Let H be a complex Hilbert space. If $A \subset B(H)$ is a unital $*$-subalgebra of $B(H)$, then the bicommutant of A coincides both with the weak and the strong closure of A in $B(H)$.*

Theorem A.1.3 (Abelian von Neumann algebras [24, Corollary III.1.5.18]). *Let N be an Abelian von Neumann algebra on a separable Hilbert space. Then there exists a standard Borel measure space (X, μ) with σ-finite measure μ such that N is isomorphic to the von Neumann algebra $L^\infty(X, \mu) \subset L^2(X, \mu)$.*

Further information on von Neumann algebras and their traces can be found, e.g., in the books by Fillmore [57] and Blackadar [24].

A.2 Weak convergence of measures

Weak convergence of measures combines topological and measure theoretic properties; for simplicity, we formulate everything for probability measures, but the same theory also applies via scaling to finite measures (with the same total measure).

Definition A.2.1 (weak convergence of measures). Let X be a separable metrisable topological space and let $\mu, \mu_0, \mu_1, \ldots$ be Borel probability measures on X. Then the sequence $(\mu_n)_{n \in \mathbb{N}}$ *weakly converges* to μ if

$$\lim_{n \to \infty} \int_X f \, d\mu_n = \int_X f \, d\mu$$

holds for all bounded continuous functions $f \colon X \longrightarrow \mathbb{C}$.

Theorem A.2.2 (portmanteau theorem [87, Theorem 17.20][21]). *Let X be a separable metrisable topological space and let $\mu, \mu_0, \mu_1, \dots$ be Borel probability measures on X. Then the following are equivalent:*

1. *The sequence $(\mu_n)_{n \in \mathbb{N}}$ converges weakly to μ.*

2. *For all continuous bounded functions $f \colon X \longrightarrow \mathbb{C}$, we have*

$$\lim_{n \to \infty} \int_X f \, d\mu_n = \int_X f \, d\mu.$$

3. *For all closed subsets $K \subset X$, we have $\limsup_{n \to \infty} \mu_n(K) \leq \mu(K)$.*

4. *For all open subsets $U \subset X$, we have $\liminf_{n \to \infty} \mu_n(U) \geq \mu(U)$.*

5. *For all μ-regular Borel sets $A \subset X$, we have $\lim_{n \to \infty} \mu_n(A) = \mu(A)$. A Borel set A is μ-regular if $\mu(\overline{A} \setminus A^\circ) = 0$.*

More information on the convergence of measures can be found in the book by Billingsley [21].

A.3 Lattices

Definition A.3.1 (lattice). Let G be a locally compact second countable topological group.

- A *lattice in G* is a discrete subgroup Γ of G with finite covolume, i.e., the measure on G/Γ induced by the Haar measure on G is finite. A lattice Γ in G is *uniform* if the quotient space G/Γ is compact.

- A lattice Γ in G is *irreducible* if for every closed non-discrete normal subgroup H of G the image of Γ in G/H under the canonical projection $G \longrightarrow G/H$ is dense.

Definition A.3.2 (semi-simple Lie group). A connected Lie group is *semi-simple* if it does not contain a non-trivial connected normal Abelian subgroup.

If G is a semi-simple Lie group, then the centre $C(G)$ of G is discrete and the quotient $G/C(G)$ is a centre-free (!) semi-simple Lie group. Every centre-free semi-simple Lie group can (essentially uniquely) be decomposed into a finite direct product of connected *simple* Lie groups.

Definition A.3.3 ($\mathbb{R}$-rank). Let G be a semi-simple Lie group that is (for some $n \in \mathbb{N}$) a closed subgroup of $\mathrm{SL}(n, \mathbb{R})$. The $\mathbb{R}$-*rank* of G is the maximal number $k \in \mathbb{N}$ such that G contains a k-dimensional Abelian subgroup that is conjugate to a subgroup of the diagonal matrices in $\mathrm{SL}(n, \mathbb{R})$. The $\mathbb{R}$-rank of G is denoted by $\mathrm{rk}_{\mathbb{R}} G$.

This definition indeed does not depend on the chosen embedding of the Lie group in question into a special linear group [79, Section 21.3].

Standard examples of lattices include the following:

lattice	ambient group	uniform?	irreducible?	$\mathbb{R}$-rank
$\mathbb{Z}^n$	$\mathbb{R}^n$	$+$	iff $n \leq 1$	n
$\mathrm{SL}(2, \mathbb{Z})$	$\mathrm{SL}(2, \mathbb{R})$	$-$	$+$	1
$\mathbb{Z} * \mathbb{Z}$	$\mathrm{SL}(2, \mathbb{R})$	$-$	$+$	1
$\mathrm{SL}(2, \mathbb{Z}) \times \mathrm{SL}(2, \mathbb{Z})$	$\mathrm{SL}(2, \mathbb{R}) \times \mathrm{SL}(2, \mathbb{R})$	$-$	$-$	2
$\mathrm{SL}(2, \mathbb{Z}[\sqrt{2}])$	$\mathrm{SL}(2, \mathbb{R}) \times \mathrm{SL}(2, \mathbb{R})$	$-$	$+$	2
$\mathrm{SL}(3, \mathbb{Z})$	$\mathrm{SL}(3, \mathbb{R})$	$-$	$+$	2

Moreover, if M is a closed Riemannian manifold, then $\pi_1(M)$ is a uniform lattice in $\mathrm{Isom}^+(\widetilde{M})$. If M is a locally symmetric space of non-compact type, then the irreducibility properties as well as the $\mathbb{R}$-rank correspond to geometric properties of M [48, Chapter 3.11].

A comprehensive introduction to lattices and their arithmeticity properties is the book by Witte Morris [136].

Bibliography

[1] M. Abért, N. Bergeron, I. Biringer, T. Gelander. Convergence of normalized Betti numbers in nonpositive curvature, preprint, 2018. arXiv:arXiv:1811.02520 [math.GT] Cited on page: 65, 70

[2] M. Abért, N. Bergeron, I. Biringer, T. Gelander, N. Nikolov, J. Raimbault, I. Samet. On the growth of L^2-invariants for sequences of lattices in Lie groups, *Ann. of Math. (2)*, 185(3), pp. 711–790, 2017. Cited on page: 60, 61, 62, 63, 65, 66, 68, 69, 70

[3] M. Abért, N. Bergeron, I. Biringer, T. Gelander, N. Nikolov, J. Raimbault, I. Samet. On the growth of L^2-invariants of locally symmetric spaces, II: exotic invariant random subgroups in rank one, preprint, 2016. arXiv:1612.09510v1 [math.GT] Cited on page: 60, 65, 70

[4] M. Abért, I. Biringer. Unimodular measures on the space of all Riemannian manifolds, preprint, 2016. arXiv:1606.03360 [math.GT] Cited on page: 70

[5] M. Abért, Y. Glasner, B. Virág. Kesten's theorem for invariant random subgroups, *Duke Math. J.*, 163(3), pp. 465–488, 2014. Cited on page: 65

[6] M. Abért, A. Jaikin-Zapirain, N. Nikolov. The rank gradient from a combinatorial viewpoint, *Groups Geom. Dyn.*, 5(2), pp. 213–230, 2011. Cited on page: 52

[7] M. Abért, N. Nikolov. Rank gradient, cost of groups and the rank versus Heegaard genus problem, *J. Eur. Math. Soc.*, 14, pp. 1657–1677, 2012. Cited on page: 48, 55

[8] M. Abért, B. Weiss. Bernoulli actions are weakly contained in any
     free action, *Ergodic Theory Dynam. Systems*, 33(2), pp. 323–333, 2013.
     Cited on page: 89

[9] I. Agol. The virtual Haken conjecture, with an appendix by I. Agol,
     D. Groves, and J. Manning, *Doc. Math.*, 18, pp. 1045–1087, 2013. Cited
     on page: 71

[10] D. Aldous, R. Lyons. Processes on unimodular random networks, *Elec-
     tron. J. Probab.*, 12(54), pp. 1454–1508, 2007. Cited on page: 66

[11] M. F. Atiyah. Elliptic operators, discrete groups and von Neumann
     algebras, *Astérisque*, 32–33, pp. 43–72, 1976. Cited on page: 24

[12] T. Austin. Rational group ring elements with kernels having irrational
     dimension, *Proc. Lond. Math. Soc.*, 107(6), pp. 1424–1448, 2013. Cited
     on page: 14

[13] G. Avramidi. Rational manifold models for duality groups, *Geom.
     Funct. Anal.*, 28(4), pp. 965–994, 2018. Cited on page: 24

[14] U. Bader, A. Furman, R. Sauer. Efficient subdivision in hyperbolic
     groups and applications, *Groups Geom. Dyn.*, 7(2), pp. 263–292, 2013.
     Cited on page: 79, 90

[15] B. Bekka, P. de la Harpe, A. Valette. *Kazhdan's Property (T)*, New
     Mathematical Monographs, 11, Cambridge University Press, 2008.
     Cited on page: 62

[16] M. B. Bekka, M. Mayer. *Ergodic theory and topological dynamics of
     group actions on homogeneous spaces*, London Mathematical Society
     Lecture Note Series, 269, Cambridge University Press, 2000. Cited on
     page: 68

[17] R. Benedetti, C. Petronio. *Lectures on Hyperbolic Geometry*, Universi-
     text, Springer, 1992. Cited on page: 74

[18] I. Benjamini, O. Schramm. Recurrence of distributional limits of finite
     planar graphs, *Electron. J. Probab.*, 6(23), 2001. Cited on page: 61

[19] N. Bergeron, D. Gaboriau. Asymptotique des nombres de Betti, in-
     variants l^2 et laminations, *Comment. Math. Helv.*, 79(2), pp. 362–395,
     2004. Cited on page: 66

[20] M. Bestvina, N. Brady. Morse theory and finiteness properties of
     groups, *Invent. Math.*, 129(3), pp. 445–470, 1997. Cited on page: 19

[21] P. Billingsley. *Convergence of Probability Measures*, second edition,
     Wiley-Interscience, 1999. Cited on page: 95

[22] I. Biringer, J. Raimbault, Ends of unimodular random manifolds, *Proc. Amer. Math. Soc.*, 145(9), pp. 4021–4029, 2017. Cited on page: 70

[23] M. Sh. Birman, M. Z. Solomjak. *Spectral Theory of Self-Adjoint Operators in Hilbert Space*, Mathematics and its Applications, 5, Springer, 1987. Cited on page: 30

[24] B. Blackadar. *Operator algebras. Theory of C^*-algebras and von Neumann algebras*, Encyclopaedia of Mathematical Sciences, 122, Operator Algebras and Non-commutative Geometry, III, Springer, 2006. Cited on page: 94

[25] A. Borel. The L^2-cohomology of negatively curved Riemannian symmetric spaces, *Ann. Acad. Sci. Fenn. Ser. A I Math.*, 10, pp. 95–105, 1985. Cited on page: 61

[26] L. Bowen. Invariant random subgroups of the free group, *Groups Geom. Dyn.*, 9(3), pp. 891–916, 2015. Cited on page: 65

[27] L. Bowen. Cheeger constants and L^2-Betti numbers, *Duke Math. J.*, 164(2), pp. 569–615, 2015. Cited on page: 65, 70

[28] S. Braun. *Simplicial Volume and Macroscopic Scalar Curvature*, PhD thesis, KIT, 2018. https://publikationen.bibliothek.kit.edu/1000086838 Cited on page: 79

[29] M. R. Bridson, M. D. E. Conder, A. W. Reid. Determining Fuchsian groups by their finite quotients, *Israel J. Math.*, 214(1), pp. 1–41, 2016. Cited on page: 89

[30] K. S. Brown. *Cohomology of Groups*, Graduate Texts in Mathematics, 82, Springer, 1982. Cited on page: 18, 19

[31] M. Bucher-Karlsson. The simplicial volume of closed manifolds covered by $\mathbb{H}^2 \times \mathbb{H}^2$, *J. Topol.*, 1(3), pp. 584–602, 2008. Cited on page: 74

[32] M. Bucher, C. Connell, J. Lafont. Vanishing simplicial volume for certain affine manifolds, *Proc. Amer. Math. Soc.*, 146, pp. 1287–1294, 2018. Cited on page: 74

[33] M. Bucher, C. Neofytidis. The simplicial volume of mapping tori of 3-manifolds, preprint, 2018. arXiv:1812.10726 [math.GT] Cited on page: 74, 79

[34] C. Campagnolo, D. Corro. Integral foliated simplicial volume and circle foliations, preprint, 2019. arXiv:1910.03071 [math.GT] Cited on page: 79

[35] C. Chabauty. Limite d'ensembles et géométrie des nombres, *Bull. Soc. Math. France*, 78, pp. 143–151, 1950. Cited on page: 64

[36] J. Cheeger, M. Gromov. L^2-cohomology and group cohomology, *Topology*, 25(2), pp. 189–215, 1986. Cited on page: 13, 24, 46

[37] I. Chifan, T. Sinclair, B. Udrea. Inner amenability for groups and central sequences in factors, *Ergodic Theory Dynam. Systems*, 36(4), pp. 1106–1129, 2016. Cited on page: 55

[38] J. M. Cohen. Zero divisors in group rings, *Comm. Algebra*, 2, pp. 1–14, 1974. Cited on page: 7

[39] C. Connell, S. Wang. Positivity of simplicial volume for nonpositively curved manifolds with a Ricci-type curvature condition, *Groups Geom. Dyn.*, 13(3), pp. 1007–1034, 2019. Cited on page: 74

[40] A. Connes, D. Shlyakhtenko. L^2-homology for von Neumann algebras, *J. Reine Angew. Math.*, 586, pp. 125–168, 2005. Cited on page: 25

[41] M. W. Davis. *The Geometry and Topology of Coxeter Groups*, London Mathematical Society Monographs, 32, Princeton University Press, 2008. Cited on page: 24

[42] D. L. DeGeorge, N. R. Wallach. Limit formulas for multiplicities in $L^2(\Gamma \setminus G)$, *Ann. of Math. (2)*, 107(1), pp. 133–150, 1978. Cited on page: 69

[43] A. Deitmar. Benjamini–Schramm and spectral convergence, *Enseign. Math.*, 64(3-4), pp. 371–394, 2018. Cited on page: 70

[44] T. Delzant. Sur l'anneau d'un groupe hyperbolique, *C. R. Acad. Sci. Paris Ser. I Math.*, 324(4), pp. 381–384, 1997. Cited on page: 7

[45] C. Deninger, K. Schmidt. Expansive algebraic actions of discrete residually finite amenable groups and their entropy, *Ergodic Theory and Dynamical Systems*, 27(3), pp. 769–786, 2007. Cited on page: 85

[46] J. Dodziuk. de Rham-Hodge theory for L^2-cohomology of infinite coverings, *Topology*, 16(2), pp. 157–165, 1977. Cited on page: 24

[47] J. Dodziuk. L^2 harmonic forms on rotationally symmetric Riemannian manifolds, *Proc. Amer. Math. Soc.*, 77(3), pp. 395–400, 1979. Cited on page: 24

[48] P. B. Eberlein. *Geometry of nonpositively curved manifolds*, Chicago Lectures in Mathematics, The University of Chicago Press, 1996. Cited on page: 96

[49] G. Elek. Betti numbers are testable, *Fete of Combinatorics and Computer Science*, pp. 139–149, Springer, 2010. Cited on page: 70

[50] M. Ershov, W. Lück. The first L^2-Betti number and approximation in arbitrary characteristic, *Documenta Math.*, 19, pp. 313–331, 2014. Cited on page: 33

[51] M. Farber. von Neumann categories and extended L^2-cohomology, *K-Theory*, 15(4), pp. 347–405, 1998. Cited on page: 13, 24

[52] D. Fauser. Integral foliated simplicial volume and S^1-actions, preprint, 2017. arXiv:1704.08538 [math.GT] Cited on page: 77, 79

[53] D. Fauser, S. Friedl, C. Löh. Integral approximation of simplicial volume of graph manifolds, *Bull. Lond. Math. Soc.*, 51(4), pp. 715–731, 2019. Cited on page: 77, 79

[54] D. Fauser, C. Löh. Variations on the theme of the uniform boundary condition, to appear in *J. Topol. Anal.*, 2020.
DOI 10.1142/S1793525320500090 Cited on page: 86

[55] D. Fauser, C. Löh, M. Moraschini, J. P. Quintanilha. Stable integral simplicial volume of 3-manifolds, preprint, 2019. arXiv:1910.06120 [math.GT] Cited on page: 76, 77, 78, 88, 90, 91

[56] J. Feldman, C. C. Moore. Ergodic equivalence relations, cohomology, and von Neumann algebras. I, *Trans. Amer. Math. Soc.*, 234(2), pp. 289–324, 1977. Cited on page: 41

[57] P. A. Fillmore. *A user's guide to operator algebras*, Canadian Mathematical Society Series of Monographs and Advanced Texts, 14, Wiley, 1996. Cited on page: 44, 94

[58] T. Finis, E. Lapid, W. Müller. Limit multiplicities for principal congruence subgroups of GL(n) and SL(n), *J. Inst. Math. Jussieu*, 14(3), pp. 589–638, 2015. Cited on page: 70

[59] S. Francaviglia, R. Frigerio, B. Martelli. Stable complexity and simplicial volume of manifolds, *J. Topol.*, 5(4), pp. 977–1010, 2012. Cited on page: 76

[60] R. Frigerio, C. Löh, C. Pagliantini, R. Sauer. Integral foliated simplicial volume of aspherical manifolds, *Israel J. Math.*, 216(2), pp. 707–751, 2016. Cited on page: 76, 77, 78, 84, 86, 87, 89

[61] A. Furman. Gromov's measure equivalence and rigidity of higher rank lattices, *Ann. of Math. (2)*, 150(3), pp. 1059–1081, 1999. Cited on page: 58

[62] A. Furman. A survey of measured group theory. In *Geometry, Rigidity, and Group Actions* (B. Farb, D. Fisher, eds.), 296–347, The University of Chicago Press, 2011. Cited on page: 1, 2, 38, 39, 40

[63] D. Gaboriau. Coût des relations d'équivalence et des groupes, *Invent. Math.*, 139(1), 41–98, 2000. Cited on page: 47, 52, 55

[64] D. Gaboriau. Invariants l^2 de relations d'équivalence et de groupes, *Publ. Math. Inst. Hautes Études Sci.*, 95, pp. 93–150, 2002. Cited on page: 25, 41, 43, 46, 52

[65] D. Gaboriau. Orbit equivalence and measured group theory, *Proceedings of the International Congress of Mathematicians. Volume III*, pp. 1501–1527, Hindustan Book Agency, 2010. Cited on page: 2, 38

[66] T. Gelander. A lecture on invariant random subgroups, *New directions in locally compact groups*, London Math. Soc. Lecture Note Ser., 447, pp. 186–204, Cambridge University Press, 2018. Cited on page: 71

[67] T. Gelander. A view on invariant random subgroups and lattices, *Proceedings of the International Congress of Mathematicians. Volume II*, pp. 1321–1344, World Sci. Publ., 2018. Cited on page: 66

[68] T. Gelander. Kazhdan–Margulis theorem for invariant random subgroups, *Adv. Math.*, 327, pp. 47–51, 2018. Cited on page: 66

[69] T. Gelander, A. Levit. Invariant random subgroups over non-archimedean local fields, *Math. Ann.*, 372(3/4), pp. 1503–1544, 2018. Cited on page: 66, 70

[70] T. Gelander, A. Levit. Local rigidity of uniform lattices, *Comment. Math. Helv.*, 93(4), pp. 781–827, 2018. Cited on page: 66

[71] E. Glasner, B. Weiss. Kazhdan's property T and the geometry of the collection of invariant measures, *Geom. Funct. Anal.*, 7(5), pp. 917–935, 1997. Cited on page: 68

[72] Ł. Grabowski. On Turing dynamical systems and the Atiyah problem, *Invent. Math.*, 198(1), pp. 27–69, 2014. Cited on page: 14

[73] M. Gromov. Volume and bounded cohomology. *Publ. Math. IHES*, 56, pp. 5–99, 1982. Cited on page: 74, 75, 79, 91, 92

[74] M. Gromov. Asymptotic invariants of infinite groups, *Geometric group theory, Vol. 2* (Sussex 1991). London Math. Soc. Lectures Notes Ser., 182, Cambridge Univ. Press, Cambridge, pp. 1–295, 1993. Cited on page: 39, 75

[75] M. Gromov. *Metric structures for Riemannian and non-Riemannian spaces*. With appendices by M. Katz, P. Pansu, and S. Semmes, translated by S. M. Bates. Progress in Mathematics, 152, Birkhäuser, 1999. Cited on page: 76, 81

[76] L. Guth. Volumes of balls in large Riemannian manifolds, *Ann. of Math. (2)*, 173(1), pp. 51–76, 2011. Cited on page: 74

[77] N. Heuer, C. Löh. The spectrum of simplicial volume, preprint, 2019. arXiv:1904.04539 [math.GT] Cited on page: 74

[78] N. Heuer, C. Löh. Transcendental simplicial volumes, preprint, 2019. arXiv:1911.06386 [math.GT] Cited on page: 74

[79] J. E. Humphreys. *Linear algebraic groups*, Graduate Texts in Mathematics, 21, Springer, 1975. Cited on page: 96

[80] T. Hutchcroft, G. Pete. Kazhdan groups have cost 1, preprint, 2018. arXiv:1810.11015 [math.GR] Cited on page: 55

[81] I. Ishan, J. Petersen, L. Ruth. Von Neumann equivalence and properly proximal groups, preprint, 2019. arXiv:1910.08682 [math.OA] Cited on page: 55

[82] N. V. Ivanov. Foundations of the theory of bounded cohomology, *J. Soviet Math.*, 37, pp. 1090–1114, 1987. Cited on page: 74, 79

[83] D.A. Kajdan. On arithmetic varieties, *Lie groups and their representations (Proc. Summer School, Bolyai János Math. Soc., Budapest, 1971)*, pp. 151–217, Halsted, 1975. Cited on page: 31

[84] H. Kammeyer. *Introduction to ℓ^2-invariants*, Springer Lecture Notes in Mathematics, 2247, 2019. Cited on page: 2, 26, 30

[85] H. Kammeyer, R. Sauer. S-arithmetic spinor groups with the same finite quotients and distinct l^2-cohomology, *Groups Geom. Dyn.*, to appear. arXiv:1804.10604 [math.GR] Cited on page: 89

[86] A. Kar, N. Nikolov. Rank gradient and cost of Artin groups and their relatives, *Groups Geom. Dyn.*, 8(2), pp. 1195–1205, 2014. Cited on page: 52

[87] A. S. Kechris. *Classical Descriptive Set Theory*, Graduate Texts in Mathematics, 156. Springer, 1995. Cited on page: 38, 49, 80, 95

[88] A. S. Kechris. Weak containment in the space of actions of a free group, *Israel J. Math.*, 189, pp. 461–507, 2012. Cited on page: 89

[89] A. S. Kechris, B. D. Miller. *Topics in Orbit Equivalence*, Springer Lecture Notes in Mathematics, 1852, 2004. Cited on page: 38, 40, 48

[90] D. Kerr, H. Li. *Ergodic theory. Independence and dichotomies*, Springer Monographs in Mathematics, Springer, 2016. Cited on page: 38, 40, 58

[91] S. Kionke. The growth of Betti numbers and approximation theorems, Borel seminar, 2017. arXiv:1709.00769 [math.AT] Cited on page: 33

[92] S. Kionke, M. Schrödl-Baumann. Equivariant Benjamini–Schramm convergence of simplicial complexes and ℓ^2-multiplicities, preprint, 2019. arXiv:1905.05658 [math.AT] Cited on page: 70

[93] M. Lackenby. Expanders, rank and graphs of groups, *Israel J. Math.*, 146, pp. 357–370, 2005. Cited on page: 33

[94] J.-F. Lafont, B. Schmidt. Simplicial volume of closed locally symmetric spaces of non-compact type, *Acta Math.*, 197(1), pp. 129–143, 2006. Cited on page: 74

[95] A. Levit. The Nevo–Zimmer intermediate factor theorem over local fields, *Geom. Dedicata*, 186, pp. 149–171, 2017. Cited on page: 66, 68

[96] G. Levitt. On the cost of generating an equivalence relation, *Ergodic Theory Dynam. Systems*, 15(6), pp. 1173–1181, 1995. Cited on page: 47

[97] C. Löh. Simplicial Volume, *Bull. Man. Atl.*, pp. 7–18, 2011 Cited on page: 74

[98] C. Löh. Rank gradient vs. stable integral simplicial volume, *Period. Math. Hung.*, 76, pp. 88–94, 2018. Cited on page: 76, 83

[99] C. Löh. Cost vs. integral foliated simplicial volume, *Groups Geom. Dyn.*, to appear. arXiv: 1809.09660 [math.GT] Cited on page: 78, 79, 84

[100] C. Löh, C. Pagliantini. Integral foliated simplicial volume of hyperbolic 3-manifolds, *Groups Geom. Dyn.*, 10(3), pp. 825–865, 2016. Cited on page: 77, 78, 79, 87, 89, 91

[101] C. Löh, R. Sauer. Bounded cohomology of amenable covers via classifying spaces, preprint, 2019. arXiv: 1910.11716 [math.AT] Cited on page: 79

[102] J. Lott. Deficiencies of lattice subgroups of Lie groups, *Bull. London Math. Soc.*, 31(2), pp. 191–195, 1999. Cited on page: 26

[103] W. Lück. Approximating L^2-invariants by their finite-dimensional analogues, *Geom. Funct. Anal.*, 4(4), pp. 455–481, 1994. Cited on page: 28

[104] W. Lück. Dimension theory of arbitrary modules over finite von Neumann algebras and L^2-Betti numbers. I. Foundations, *J. Reine Angew. Math.*, 495, pp. 135–162, 1998. Cited on page: 13, 14, 24

[105] W. Lück. *L^2-Invariants: Theory and Applications to Geometry and K-Theory*, Ergebnisse der Mathematik und ihrer Grenzgebiete, 44, Springer, 2002. Cited on page: 2, 7, 10, 13, 14, 15, 20, 22, 24, 35, 45, 61, 79, 82, 91

[106] W. Lück. Approximating L^2-invariants by their classical counterparts, *EMS Surveys in Math. Sci.*, 3(2), pp. 259–344, 2016. Cited on page: 33, 34

[107] A. Malcev. On isomorphic matrix representations of infinite groups, *Rec. Math. [Mat. Sbornik] N.S.*, 8(50), 405–422, 1940. Cited on page: 28

[108] G. A. Margulis. *Discrete subgroups of semisimple Lie groups*, Ergebnisse der Mathematik und ihrer Grenzgebiete, 17, Springer, 1991. Cited on page: 62

[109] N. Monod, Y. Shalom. Orbit equivalence rigidity and bounded cohomology, *Ann. of Math. (2)*, 164(3), pp. 825–878, 2006. Cited on page: 57

[110] H. J. Munkholm. Simplices of maximal volume in hyperbolic space, Gromov's norm, and Gromov's proof of Mostow's rigidity theorem (following Thurston). In *Topology Symposium, Siegen 1979*, Lecture Notes in Mathematics, 788, pp. 109–124. Springer, 1980. Cited on page: 74

[111] H. Namazi, P. Pankka, J. Souto. Distributional limits of Riemannian manifolds and graphs with sublinear genus growth, *Geom. Funct. Anal.*, 24(1), pp. 322–359, 2014. Cited on page: 70

[112] S. Neshveyev, S. Rustad. On the definition of L^2-Betti numbers of equivalence relations, *Internat. J. Algebra Comput.*, 19(3), pp. 383–396, 2009. Cited on page: 41

[113] B. Nica. Linear groups – Malcev's theorem and Selberg's lemma, preprint, 2013. arXiv:1306.2385 [math.GR] Cited on page: 28

[114] D. S. Ornstein, B. Weiss. Ergodic theory of amenable group actions. I. The Rohlin lemma, *Bull. Amer. Math. Soc.*, 2(1), pp. 161–164, 1980. Cited on page: 40

[115] P. Papasoglu. Homogeneous trees are bi-Lipschitz equivalent, *Geom. Dedicata*, 54, 301–306, 1995. Cited on page: 26

[116] N. Pappas. Rank gradient and p-gradient of amalgamated free products and HNN extensions, *Comm. Algebra*, 43(10), pp. 4515–4527, 2015. Cited on page: 52

[117] D. S. Passman. *Group rings, crossed products and Galois theory*, CBMS Regional Conference Series in Mathematics, 64, AMS, 1986. Cited on page: 7

[118] A. L. T. Paterson. *Amenability*, Mathematical Surveys and Monographs, 29, AMS, 1988. Cited on page: 40

[119] H. D. Petersen. L^2-Betti numbers of locally compact groups, *C. R. Math. Acad. Sci. Paris*, 351(9–10), pp. 339–342, 2013. Cited on page: 25, 47, 69

[120] H. D. Petersen, A. Valette. L^2-Betti numbers and Plancherel measures, *J. of Funct. Analysis*, 266(5), pp. 3156–3169, 2014. Cited on page: 69

[121] S. Popa, S. Vaes. Vanishing of the first continuous L^2-cohomology for II_1 factors, *Int. Math. Res. Not.*, 2015(12), pp. 3899–3907, 2015. Cited on page: 25

[122] V. Runde. *Amenability*, Springer Lecture Notes in Mathematics, 1774, Springer, 2002. Cited on page: 40

[123] R. Sauer. L^2-Betti numbers of discrete measured groupoids, *Internat. J. Algebra Comput.*, 15(5–6), pp. 1169–1188, 2005. Cited on page: 14, 41, 43, 44, 45

[124] R. Sauer. Amenable covers, volume and L^2-Betti numbers of aspherical manifolds, *J. Reine Angew. Math.*, 636, 47–92, 2009. Cited on page: 79, 83, 86

[125] R. Sauer. Volume and homology growth of aspherical manifolds, *Geom. Topol.*, 20, 1035–1059, 2016. Cited on page: 76

[126] M. Schmidt. *L^2-Betti Numbers of $\mathcal{R}$-spaces and the Integral Foliated Simplicial Volume.* PhD thesis, Westfälische Wilhelms-Universität Münster, 2005. http://nbn-resolving.de/urn:nbn:de:hbz:6-05699458563 Cited on page: 38, 77, 78, 82

[127] M. Schrödl-Baumann. ℓ^2-Betti numbers of random rooted simplicial complexes, to appear in *Manuscripta math.*, 2020. DOI 10.1007/s00229-019-01131-y Cited on page: 70

[128] T. Soma. The Gromov invariant of links, *Invent. Math.*, 64, pp. 445–454, 1981. Cited on page: 75, 90

[129] C. Strohm (= C. Löh). *The Proportionality Principle of Simplicial Volume.* Diplomarbeit, WWU Münster, 2004. arXiv:math.AT/0504106 Cited on page: 74

[130] G. Stuck, R. Zimmer. Stabilizers for ergodic actions of higher rank semisimple groups, *Ann. of Math. (2)*, 139(3), pp. 723–747, 1994. Cited on page: 68

[131] A. Thom. A note on normal generation and generation of groups, *Communications in Mathematics*, 23(1), pp. 1–11, 2015. Cited on page: 34

[132] W. P. Thurston. *The Geometry and Topology of 3-Manifolds*, mimeographed notes, 1979. http://www.msri.org/publications/books/gt3m. Cited on page: 74

[133] R. Tucker-Drob. Invariant means and the structure of inner amenable groups, preprint, 2014. arXiv:1407.7474 [math.GR] Cited on page: 55

[134] A. M. Vershik. Totally nonfree actions and the infinite symmetric group, *Mosc. Math. J.*, 12(1), pp. 193–212, 216, 2012. Cited on page: 65

[135] K. Whyte. Amenability, bi-Lipschitz equivalence, and the von Neumann conjecture, *Duke Math. J.*, 99(1), 93–112, 1999. Cited on page: 26

[136] D. Witte Morris. *Introduction to Arithmetic Groups*, Deductive Press, 2015. Cited on page: 96

[137] K. Yano. Gromov invariant and S^1-actions, *J. Fac. Sci. U. Tokyo, Sec. 1A Math.*, 29(3), pp. 493–501, 1982. Cited on page: 74, 79

[138] R. J. Zimmer. *Ergodic theory and semisimple groups*. Monographs in Mathematics, 81. Birkhäuser Verlag, 1984. Cited on page: 65

Symbols

Symbols

1	the trivial group,
$\lvert \cdot \rvert$	cardinality, absolute value,
$\lVert \cdot \rVert$	operator norm,
$\langle \cdot, \cdot \rangle$	inner product,
$\cap$	intersection of sets,
$\cup$	union of sets,
$\sqcup$	disjoint union of sets,
$\subset$	subset relation (equality is permitted),
$\lVert M \rVert$	simplicial volume of M, 74
$\lVert M \rVert_{\mathbb{Z}}$	integral simplicial volume of M, 75
$\lVert M \rVert_{\mathbb{Z}}^{\infty}$	stable integral simplicial volume of M, 75
$\lvert M \rvert$	integral foliated simplicial volume, 77
$\lvert M \rvert^{\alpha}$	parametrised simplicial volume of M with respect to the standard action α, 77
$\sim_{\mathrm{ME}}$	measure equivalent, 39
$\sim_{\mathrm{OE}}$	orbit equivalent, 39
$\sim_{\mathrm{SOE}}$	stably orbit equivalent, 39
$\times$	cartesian product,
$\cdot^{\times}$	set of units,

B

$b_n^{(2)}$	n-th L^2-Betti number, 19, 20, 43
$B(\ell^2 \Gamma)$	algebra of bounded operators on $\ell^2 \Gamma$, 9

C

$\mathbb{C}$	set of complex numbers,
$C_*^{(2)}$	L^2-chain complex, 19
$\mathbb{C}\Gamma$	complex group ring of Γ, 6
χ	Euler characteristic, 22
χ_A	characteristic function,
cost	cost, 47
$\mathbb{C}\mathcal{R}$	equivalence relation ring, 42
C_*^{sing}	singular chain complex,

D

d minimal rank, 33

$\dim_{N\Gamma}$ von Neumann dimension over $N\Gamma$, 11

$\dim_{N\mathcal{R}}$ von Neumann dimension over $N\mathcal{R}$, 42

$\partial_S F$ S-boundary of a set F, 85

E

e the neutral element,

F

F_n free group of rank n,

frk   fundamental rank, 61

G

$\widehat{\Gamma}$ profinite completion of Γ, 38

H

$\mathbb{H}^n$ hyperbolic n-space,

H_* (singular/cellular) homology,

$H_*^{(2)}$ (reduced) L^2-homology, 19

I

$\mathrm{IRS}(G)$ space of all invariant random subgroups of G, 65

L

$\ell^2\Gamma$ ℓ^2-algebra of Γ (over $\mathbb{C}$), 8

N

$\mathbb{N}$ set of natural numbers: $\{0, 1, 2, \dots\}$,

$N\Gamma$ group von Neumann algebra of Γ, 9

P

$N\mathcal{R}$ von Neumann algebra of the relation $\mathcal{R}$, 42

$\mathrm{pdim}_{N\Gamma}$ von Neumann dimension over $N\Gamma$ for projectives, 13

$\mathrm{pdim}_{N\mathcal{R}}$ von Neumann dimension over $N\mathcal{R}$ for projectives, 43

Q

$\mathbb{Q}$ set of rational numbers,

R

$\mathbb{R}$ set of real numbers,

rg   rank gradient, 33

$\mathcal{R}_{\Gamma \curvearrowright X}$ orbit relation of $\Gamma \curvearrowright X$, 41

$\mathrm{rk}_{\mathbb{R}}$ real rank, 61

S

Σ_g oriented closed connected surface of genus g,

$\mathrm{Sub}(G)$ space of closed subgroups of G, 64

T

tr_Γ von Neumann trace, 9, 10

$\mathrm{tr}_\mathcal{R}$ von Neumann trace, 42

V

v_3 volume of ideal regular simplices in $\mathbb{H}^3$, 87

Z

$\mathbb{Z}$ set of integers,

$Z(M; R)$ set of R-fundamental cycles of M, 74

Index

A

absolute rank gradient, 33
amenable, 40
    L^2-Betti number, 46
    dynamical characterisation, 40
    Følner sequence, 85
    inner amenable, 55
    integral foliated simplicial volume, 86
    simplicial volume, 74, 84
approximation theorem, 28, 29, 60, 61
    lattices, 60, 61
    simplicial volume, 90
Atiyah conjecture, 14

B

Benjamini–Schramm convergence, 61, 66
Bernoulli shift, 38
    weak containment, 88
Betti number estimate, 81, 82
Betti number gradient, 33, 81, 87
bicommutant theorem, 94
BS-approximation for lattices, 60, 61
BS-convergence, 61, 66

C

Chabauty topology, 64
conjecture
    Atiyah, 14
    fixed price, 55
    Gromov, 75
    Kaplansky, 7
    Singer, 24
cost, 47
    finite group, 58
    fixed price, 55
    inner amenable, 55
    L^2-Betti number estimate, 52
    of $\mathbb{Z}$, 58
    profinite completion, 48
    property (T), 55
    simplicial volume, 84
cost estimate, 52, 84
crossed product, 43
CW-complex
    finite type, 18

D

deficiency, 26
dimension
    via trace, 9
    von Neumann, 11, 13, 42

© The Author(s), under exclusive license to Springer Nature Switzerland AG 2020
C. Löh, *Ergodic Theoretic Methods in Group Homology*,
SpringerBriefs in Mathematics, https://doi.org/10.1007/978-3-030-44220-0

dynamical characterisation of amen-
        able groups, 40
dynamical view, 37, 77

E

EMD*, 89
equivalence relation
        cost, 47
        graphing, 47
        L^2-Betti number, 43
        measured standard, 41
        ring, 42
        standard, 41
        von Neumann algebra, 42
ergodic, 38
essentially free, 38
Euler characteristic, 22, 75
extended von Neumann dimension,
        13

F

faithfulness, 9
Feldman–Moore theorem, 41
finite type, 18
        Γ-CW-complex, 18
        CW-complex, 18
        group, 18
finite von Neumann algebra, 44
fixed price problem, 55
Følner sequence, 85
free Γ-CW-complex, 18
        finite type, 18
free group, 23, 35, 46, 92
fundamental rank, 61

G

Γ-CW-complex, 18
gradient invariant, 32, 34, 87
graphing, 47
        cost, 47
Gromov norm, *see* simplicial vol-
        ume
group
        amenable, 40, 85
        cost, 47

deficiency, 26
finite type, 18
Heisenberg, 57
Hopfian, 35
inner amenable, 55
L^2-Betti number, 20
property (T), 55
property (T), 62
rank gradient, 33
residually finite, 28, 38, 57
group ring, 6
        universal property, 7
group von Neumann algebra, 9

H

Heisenberg group, 57
Hilbert module, 8, 15
        morphism, 8
        weak isomorphism, 12
homological gradient invariant, 32,
        87
homotopy invariance, 20
Hopfian, 35
hyperbolic manifold, 24, 63, 71, 87
        simplicial volume, 74, 75, 78,
        87
        stable integral simplicial vol-
        ume, 87

I

index
        of an ME coupling, 39
        of an SOE, 39
inner amenable, 55
integral foliated simplicial volume,
        77
        amenable, 86
        aspherical 3-manifold, 90
        cost estimate, 84
        gradient invariants, 87
        L^2-Betti number estimate, 82
        locally symmetric space, 89
        profinite completion, 79
        proportionality principle, 88,
        89

weak containment, 88
invariant random subgroup, 59, 65
irreducible lattice, 95
IRS, *see* invariant random subgroup

K

Kaplansky conjecture, 7
Kazhdan inequality, 31
Kazhdan's property (T), 62
Künneth formula, 21

L

L^2-Betti number
    amenable groups, 46
    approximation, 28, 60, 61
    cost, 52
    degree 0, 21
    equivalence relations, 43
    generalisations, 24
    groups, 20
    homotopy invariance, 20
    inner amenable, 55
    integral foliated simplicial volume, 82
    Künneth formula, 21
    locally symmetric space, 61
    Morse inequality, 58
    OE-invariant, 46
    Poincaré duality, 21
    property (T), 55
    proportionality principle, 46
    QI?, 26
    rank gradient, 34
    restriction formula, 21, 43
    self-map, 35
    simple examples, 23
    simplicial volume, 75
    spaces, 19, 20
    stable integral simplicial volume, 81
    topological group, 46, 57
L^2-chain complex, 19
L^2-homology, 19
Laplacian, 35
lattice, 39, 40, 95

irreducible, 95
locally symmetric space, 61, 89
Lück approximation, 28

M

measure equivalence, 39, 40
measured group theory, 38
measured standard equivalence relation, *see* equivalence relation
ME coupling, 39
    cocycle, 58
    index, 39
Morse inequality, 58

N

Nevo–Stück–Zimmer theorem, 68

O

orbit equivalence, 39, 40
orbit relation, 41
Ornstein–Weiss theorem, 40

P

parametrised simplicial volume, 77
Plancherel measure, 69
Poincaré duality, 21, 81, 82
polar decomposition, 13
portmanteau theorem, 31, 95
positivity, 9
profinite completion, 38, 57
    cost, 48
    simplicial volume, 79, 89
property (T), 62
    cost, 55
proportionality principle
    L^2-Betti number, 46
    simplicial volume, 74, 88, 89

Q

QI-invariance, 26

R

$\mathbb{R}$-rank, 96

rank gradient, 33
  of products, 35
  stable integral simplicial vol-
      ume, 83
  via cost, 48, 56
reduced L^2-homology, 19
residual chain, 28
residually finite group, 28, 38, 57
residually finite view, 28, 75
restriction formula, 12, 21, 43
Rokhlin lemma, 86

S

S-boundary, 85
self-map, 35
semi-simple, 95
seven samurai, 60
simplicial volume, 73, 74
  amenable, 74
  cost, 79
  hyperbolic manifold, 74
  integral foliated, 77
  L^2-Betti number, 75
  low dimensions, 75
  multiplicativity, 74, 92
  parametrised, 77
  sphere, 92
  stable integral, 75
  surface, 92
  torus, 92
Singer conjecture, 24, 92
spectral measure, 30
stable integral simplicial volume,
      75
  amenable, 84
  hyperbolic manifold, 87
  L^2-Betti number, 81
  rank gradient, 83
stable orbit equivalence, 39, 40
  index, 39
standard action, 38
  ergodic, 38
  essentially free, 38
standard Borel space, 38

standard equivalence relation, *see*
      equivalence relation
surface
  simplicial volume, 92
surface group, 23, 35

T

theorem
  approximation, 28, 29, 60, 61
  bicommutant, 94
  BS-approximation, 60, 61
  Feldman–Moore, 41
  Lück approximation, 28
  Nevo–Stück–Zimmer, 68
  Ornstein–Weiss, 40
  portmanteau, 31, 95
  Rokhlin lemma, 86
thin part, 61
trace
  von Neumann, 9, 10
trace property, 9

U

uniform discreteness, 62
universal property
  group ring, 7

V

view
  dynamical, 37, 77
  residually finite, 28, 75
von Neumann algebra, 94
  finite, 44
  of a group, 9
  of a measured equivalence re-
      lation, 42
von Neumann dimension, 11, 13,
      42
von Neumann trace, 9, 10, 42

W

weak containment, 88
weak convergence, 30, 35, 94
weak isomorphism, 12
weakly exact, 12

Printed in the United States
By Bookmasters